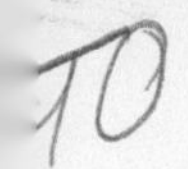

6-6-74

350 ways to SAVE ENERGY (AND MONEY) IN YOUR HOME AND CAR

350 ways to SAVE ENERGY (AND MONEY)

IN YOUR HOME AND CAR

BY

HENRY R. SPIES, SEICHI KONZO, JEAN CALVIN,
AND WAYNE THOMS

Preface by John C. Sawhill

Administrator of the Federal Energy Office

Drawings by Carl J. De Grootne

CROWN PUBLISHERS, INC., NEW YORK

Library of Congress Catalog Card Number: 74–79496

Printed in the United States of America
Published simultaneously in Canada by General Publishing Company Limited

Design: Kendra McKenzie

Contents

Preface

Americans traditionally respond actively to crisis primarily because they view any problem as a challenge. The challenge is explored and discussed, solutions—sometimes ingenious—are devised, and alternatives examined and tested until a final resolution occurs. This problem-solving tradition, along with all the ingenuity and determination of the American public, must be brought to bear on one of the most serious problems facing our nation—energy.

The energy situation is knotty, complicated by intricate industry structures, exotic technology, and enormous social and economic implications. If, however, one were to reduce the energy problem to its simplest terms, it might be put this way: the United States demand for energy is growing at a faster rate than the United States supply of energy.

We are working hard to find ways to increase the supply of energy. Intensive research and development efforts are under way to improve existing technologies and find new feasible sources of energy. But solutions on the supply side of the problem are future solutions, five, ten, twenty, or even fifty years in the future, and the problem is now.

What we can do right now to meet the energy challenge is to tighten our belts in terms of energy use. We must seek to reduce our demand for energy by using more frugally the energy that is presently available to us. Energy conservation does not come easily. It is not always convenient to bicycle to the store, to arrange car pools, or to turn down our furnaces to 68 degrees in the winter. Changes in energy consumption patterns require modifications of life-styles, and these modifications come slowly.

Nevertheless, we are optimistic that Americans will make a conscious choice for energy-saving life-styles once they understand the necessity for and methods of energy conservation. The why will be addressed elsewhere—this book tells how.

—John C. Sawhill
Administrator of the
Federal Energy Office
Washington, D.C.

PART ONE

THE HOME

1

What You Should Know about Heat and Cold

WHY YOU SHOULD BE CONCERNED

Heat and cold are what the first part of this book is all about. We are warm bodies generating heat by consuming food—we can exist only as long as we are in thermal balance with our environment. For this purpose we have built shelters, which we call homes, residences, or apartments. The entire purpose of comfort air conditioning (in winter as well as summer) is to produce a thermal environment that provides a balance between heat generation and heat dissipation *without our being conscious of the environment.*

Now we are being called upon to produce an indoor thermal environment with maximum efficiency, or at least up to the limit of our ability to pay for this efficiency. At the same time we are called upon to suddenly change the environmental level to that which we considered normal some fifty years ago. In other words, we are called upon to be comfortable at 68 degrees Fahrenheit, when we have become accustomed to a year-round environmental temperature of 75 degrees. This is not easy to do without some changes in our life-style and in our shelter. The purpose of the first part of this book is to emphasize the many different measures that can be taken to live within our energy budget *without greatly sacrificing the thermal comfort standards that we have attained in this country and in Canada.*

Continental North America offers many climatic extremes—wide variations in temperature and humidity undreamed of in many other countries. In this rugged climate we need to build more protective homes than most Europeans would ever consider necessary. A large part of the populated sections of the United States can be subjected

to 95 degree temperatures in the summer and sub-zero temperatures in the winter. We were forced to build better heating and cooling systems for homes—and we have. Residential heating and cooling systems of top quality are available to most Americans.

In considering a temperature control system many people fail to think of insulation and storm sash, which are as important or more important than other more visible or decorative home features. We believe this attitude must be changed. An important question the home buyer must ask is: *"How much does it cost to heat and cool the house?"* The seller should be expected to offer proof of actual temperature control costs by showing past bills for electricity, gas, and oil.

One thing should be apparent to everyone. Those who pay the highest heating and cooling bills are probably not getting the kind of thermal comfort conditions that we are capable of producing. If you pay twice as much as your neighbor for heating the same size house, you are not only paying more, but probably getting less.

This presents a problem of great magnitude for the responsible authorities in this country. Gas, oil, and electrical costs are increasing and will undoubtedly continue to increase. Those who can afford to make even some of the improvements that are suggested here will enjoy increased thermal comfort without losing the battle of rising fuel costs.

Those individuals who are barely able to make ends meet with existing prices are faced with a dilemma: Housing improvements cost money, and in order to stay even with rising energy costs it is necessary to invest in home improvement. Perhaps a social agency can find some solution to this predicament. We will point out the reasons for certain necessary changes in heating systems and will try to show how much fuel can be saved by home improvements. The rest is up to you —owners and tenants.

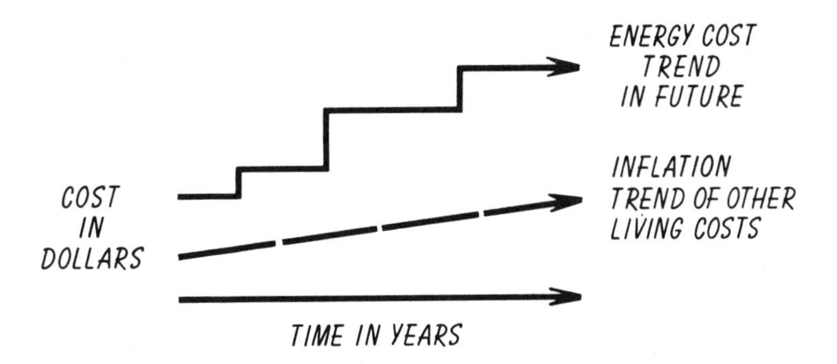

WHAT SOME OF THE TERMS MEAN

Some unfamiliar and technical terms are necessary to the explanations that appear in some sections of this book. Many are so seldom used in ordinary conversation and reading that brief definitions are necessary. We will use a layman's definition of terms and will avoid fine technical distinctions.

Heat Units

We need to provide comparative numbers on many items that will be discussed so the units of measurement of temperature and heat must be defined.

British Thermal Unit. Btu stands for a unit of thermal energy equal to the amount of heat required to raise the temperature of 1 pound of water 1 degree Fahrenheit at or near 39.2 degrees Fahrenheit, its temperature of maximum density. In other words, 5 pounds of water in a kettle raised 160 degrees requires 5 x 160 or 800 Btu. As an approximate measure, 1 Btu corresponds to the heat given off by burning 1 wooden kitchen match.

Btu Per Hour. This term is commonly abbreviated Btuh and is the *rate* of heat generation. The time unit used here is commonly 1 hour. For example, a boiler may be rated to produce 150,000 Btuh, or a window air conditioner may have a rated capacity of 12,000 Btuh.

Degree Fahrenheit. The temperatures used here are in terms of degrees Fahrenheit, the usual scale in the United States. On this temperature scale, water boils at 212 degrees at sea-level pressure and ice melts at 32 degrees.

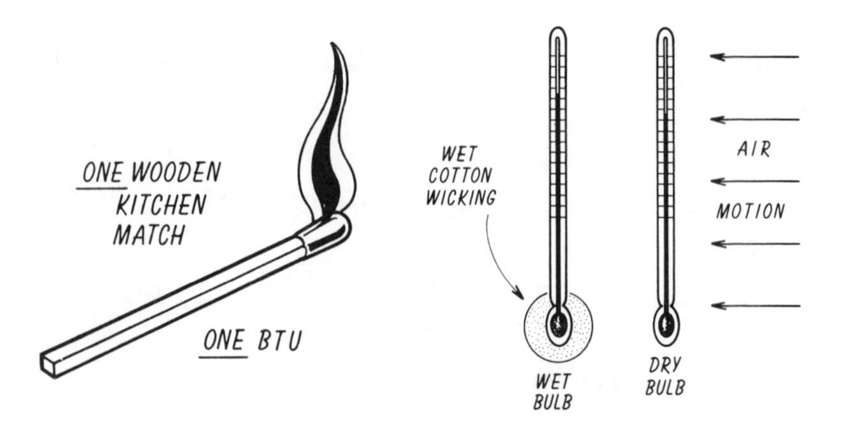

Air and Vapor Properties

The air that surrounds us is a mixture of gases, mostly nitrogen and oxygen, with varying amounts of water vapor. The latter component, water vapor, plays a most important role in daily living and in house construction and maintenance, so that some understanding of the properties of air and water vapor will be helpful.

Dry-Bulb Temperature (*db*). The ordinary glass thermometer provides us with the actual dry-bulb temperature of the air. The term is used because the thermometer bulb (usually containing mercury or alcohol) is dry and not moist.

Wet-Bulb Temperature (*wb*). A wet-bulb temperature reading is obtained when the thermometer bulb is encased in a wicking that has been completely saturated with water. The evaporation of water from the wick produces a cooling effect and the wet-bulb temperature is usually lower than the dry-bulb temperature. The temperature indicated by a wet-bulb thermometer is lower than the actual air temperature. The difference between the dry-bulb and wet-bulb temperatures is related to the relative humidity of the air.

Relative Humidity (*rh*). This is a measurement, expressed as a percentage, which indicates the amount of water vapor in the air compared to the amount that the air could contain if it were completely saturated with moisture. For example, air at 68 degrees can hold as much as 0.0147 pounds of water for each pound of air when it is completely saturated. If we attempted to add more water vapor into this saturated air, the moisture would appear as a fog and eventually fall out of the air. On the other hand, air at 40 degrees can hold only 0.0052 pounds of water vapor for each pound of air when it is completely saturated or only about one-third as much air at 68 degrees. Yet both conditions would be described as 100 percent relative humidity for the temperatures involved.

When 40 degree outdoor air that is completely saturated enters the house through infiltration, the air becomes warm and reaches 68 degrees. However, if we assume that we have not added any other moisture to the air, this indoor air now shows a relative humidity of only 35 percent at the new temperature. This always happens: When cold outdoor air leaks into a heated building, the new relative humidity will be lower than the original relative humidity of the outdoor air.

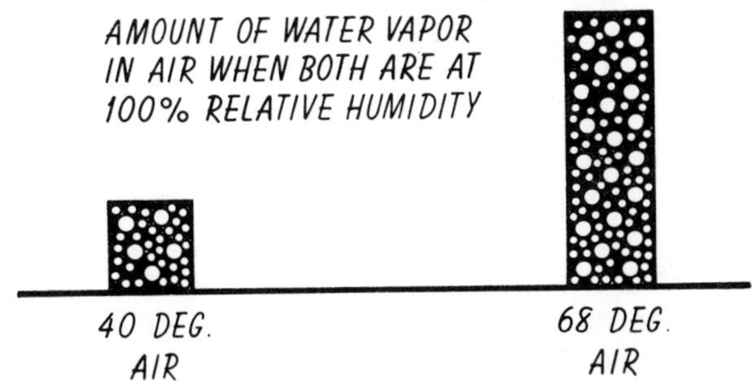

Dew-Point Temperature (dp). This is the temperature at which the water vapor in a volume of air just starts to form visible condensation. For example, a window surface exposed to cold outdoor air first shows signs of moisture condensation when the glass surface drops to the dew-point temperature of the room air. Room air at 68 degrees and 40 percent relative humidity has a dew-point temperature of 42.5 degrees. This means that when the indoor surface of the glass drops to this value, the windows will start to fog.

Indoor and Outdoor Temperatures

The heating and air-conditioning engineer is concerned about temperature, humidity, dust content, bacterial content, air motion, noise, surface temperatures, and air distribution for any space. Of all these items, the temperature of the air is the most important factor.

Indoor Design Temperature. Heating engineers select an indoor air temperature when determining the heat loss of a building. The design

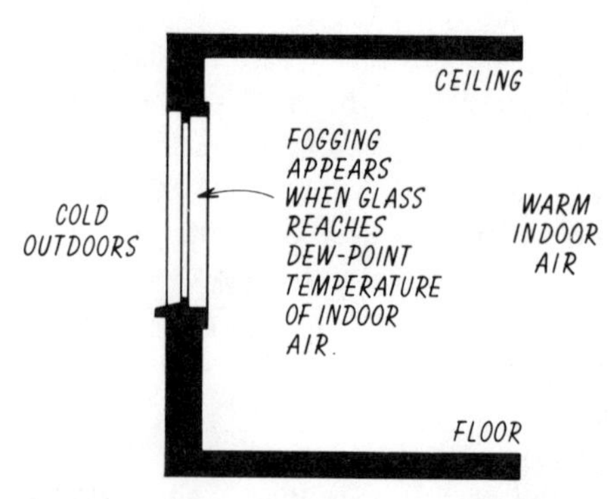

heat loss is useful in selecting the capacity of heating equipment for a particular room or an entire building. In the 1920s the indoor design temperature was 70 degrees; currently, a value of 75 degrees has been common. If the actual operating temperature is held at 68 degrees, the equipment selected in the past will be slightly oversized for current requirements. Time will tell whether the design temperature for the future will revert back to 70 degrees or be reduced to 68 degrees.

Outdoor Air Temperature. A typical variation in daily outdoor temperature shows two extreme values. The outdoor *minimum* temperature normally occurs just before sunrise, after a long period when the earth has radiated heat into outer space. The outdoor *maximum* temperature occurs around 3:00 to 4:00 P.M., several hours after the sun has reached its zenith and the earth has been receiving radiant heat for many hours. On any given day temperature variations will not correspond exactly to this pattern, but it does accurately represent what occurs when other factors don't intervene. Sudden wind shifts or changes in cloud cover can affect the outdoor temperature on a particular day.

For any given locality the heating engineer selects an outdoor design temperature for the winter heating season. This design temperature is normally about 10 to 15 degrees above the coldest temperature recorded for that locality. For example, a design value of −20 degrees is common along the Canadian border in the Midwest and a value of +20 degrees is common in the Gulf states. Installers of furnaces and boilers will be able to supply you with local outdoor design temperatures.

Design Temperature Difference. This is the difference in temperature

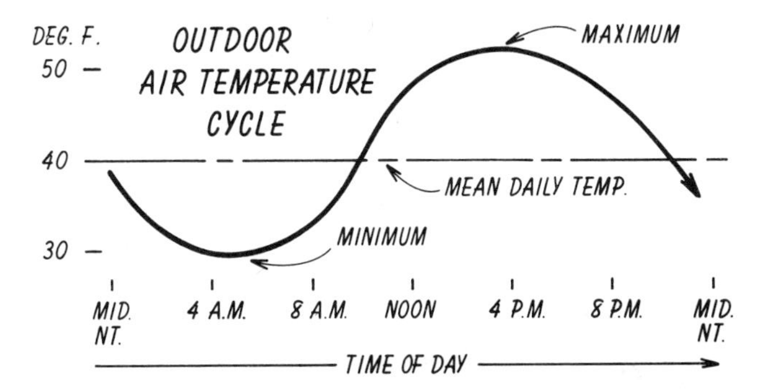

between the indoor design temperature and the outdoor design temperature for a locality. A design temperature difference of 75 degrees is a representative value for large sections of the populated areas of the Midwest and eastern parts of the United States. The range may be as much as 95 degrees in some northern cities and as little as 45 degrees in warmer climates.

All heating equipment designed and installed during the past twenty years or so may have a slightly greater capacity than will be necessary in the future.

For example, if future design temperature differences are based on an indoor temperature of 68 degrees in place of the current 75 degrees, then present design temperature differences would be reduced 7 degrees. This corresponds to a reduction of about 8 percent in northern climates and as much as 15 percent in mild climates. Savings in installed equipment, as well as operating costs, will be possible with these revised design standards. These savings have not yet taken place, but they may once the entire country is geared to lower indoor air temperatures.

Degree Days (*DD*). The average of the daily minimum and maximum temperatures for any given day is used in determining degree days. This term was first used by George Segeler about fifty years ago and is still useful as a measure of heating requirements. He discovered that steam for heating was not required when the mean daily temperature was 65 degrees or warmer. (The mean daily temperature is the average of the minimum and maximum outdoor air temperatures during a 24-hour period.) He also found that the steam requirements increased proportionally as the mean daily temperature became lower.

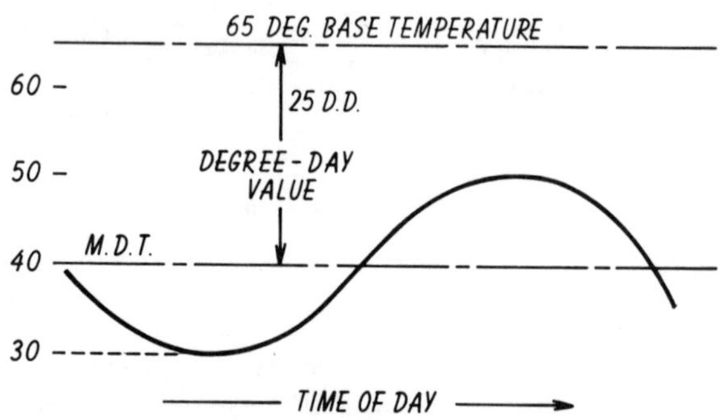

The degree-day value *for any given day,* therefore, is the difference between 65 degrees and the mean daily temperature for the day. For example, if the mean daily temperature is 40 degrees, the degree days are 65 minus 40 or 25 degree days. The colder the day, the larger is the degree-day number for that day. If the outdoor air is warm and above 65 degrees, the degree days are not counted. There are few degree days in the early part of September, but in the coldest part of winter the degree days become quite large. For example, when the mean daily temperature is 0 degrees, the degree days amount to 65 for that single day.

On September 1 most weather service stations report the number of degree days that have accumulated up to that date. For example, a city weather station may report that 1,500 degree days have accumulated by December 15. Frequently, the report will also indicate that the normal accumulation for that date is about 1,850 degree days, and that the current season has been warmer than normal *up to that time.* Or the report might indicate that last year the value on the same date was 2,000 degree days, and that the current season should have required about 25 percent less fuel than was consumed last year up to that date. (Note that these numbers give no indication of what might take place in the future.)

For every major city in the United States and Canada the seasonal total for degree days is tabulated by the National Weather Service. In the heavily populated cities extending near 40 degrees north latitude, the seasonal totals range from 5,000 to 6,500 degree days (Chicago—6,300; Cleveland—5,700; Pittsburgh—5,100; Philadelphia—4,500; New York City—5,100; Boston—5,800). In colder regions the seasonal totals will be higher; for example, the areas around Hudson Bay will show totals of over 15,000 degree days. On the other hand, New Orleans shows a little over 1,100, Los Angeles about 1,400, and Miami less than 200. The seasonal degree-day values are obtainable from local weather stations. Normally, variations of as much as 10 percent from the norm can be expected for any given season.

The degree day will become more prominent in the news as the fuel policies of the nation are changed to take into account geographical weather variations. Currently, fuel suppliers make use of degree-day data to handle consumer fuel needs. On a national scale the degree-day data will be used to move more oil into those regions where the current season shows the greatest need.

If the future standard for indoor temperature is set for 68 degrees, it is possible that a new base temperature will have to be established for determining degree days. This is especially true if more houses are better insulated. This could mean a new set of smaller degree-day accumulations for every city in this country. This is an engineering problem, but it will affect fuel allocations for individual houses and for regions, if a system of fuel rationing is ever put into effect.

Ventilation and Infiltration

In the past it was thought that if we did not open windows and allow "fresh air" to come into the house, we would suffer from carbon dioxide accumulations produced by the respiration of human beings and animals and by the combustion products of open gas flames. This theory has been proven completely false for the American home. The normal leakage of air into a structure is many times greater than is needed for minimum living requirements. The only reason for permitting any fresh air to enter a house is to keep odors down.

Air Motion. The speed of air currents in the room is expressed in feet per minute. For example, the rising convection currents from a cigarette will have a velocity of about 30 to 40 feet per minute. Air motion in excess of about 50 feet per minute within the home is considered to be excessive. We would consider air moving at such a speed a "draft." Air motion of less than about 30 feet per minute is sometimes referred to as "stagnant air." This can be seen when cigarette smoke hangs in the room like a stratified layer.

Air Changes. The air in a room can be changed by supplying air through a register and removing it through a return-air grille. The number of times that the air in a room is completely replaced is referred to as the number of "air changes per hour." For example, a central air-cooling unit will provide from 4 to 8 air changes per hour; a forced-air heating system (without cooling coils) will show about half that amount. There is no advantage in having a large number of air changes; in general, poorly insulated houses require more air changes than well-insulated houses.

Infiltration. Outdoor air will leak into a house and displace conditioned air at some other part of the house. This air leakage is referred to as infiltration air, and occurs mainly at cracks around doors, windows, and through walls. The infiltration rate increases sharply as the wind velocity increases, and also as the indoor-outdoor temperature difference becomes larger. The infiltration rate may vary from ¼ air change per hour in a well-built house during the summer to 1½ air changes per hour in a loosely built house in the winter. (The air changes given for air leakage are not the same as the air changes mentioned above for forced circulation within the room.)

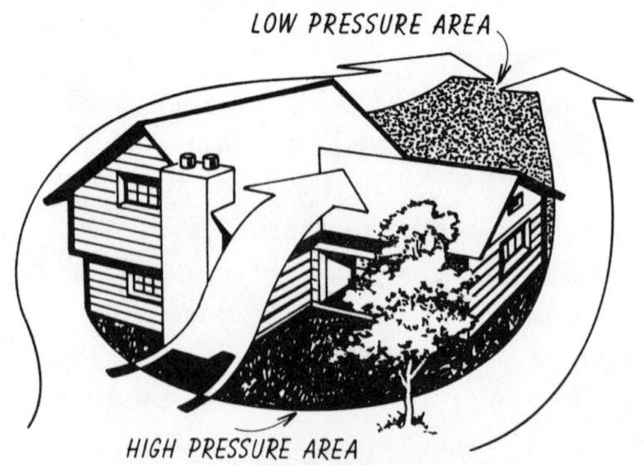

A HIGH PRESSURE AREA IS CREATED WHEN AIR STRIKES A BUILDING. LOW PRESSURE AREAS ARE CREATED AS AIR MOVES AROUND A BUILDING.

Ventilation. The intentional replacement of indoor air with outdoor air by means of a fan is called ventilation. In this case the amount of outdoor air brought into a building is under positive control. In commercial buildings a ventilating fan brings outdoor air into the structure where it is conditioned before being released into the occupied space.

In the case of homes, this is rarely done. However, ordinary kitchen and bathroom ventilating fans do discharge room air to the outdoors, and the house is placed under a slight suction. This lower indoor pressure will increase the leakage of outdoor air into the structure through infiltration. Basically, any air that leaks or is discharged outward will be replaced by an incoming air supply from outdoors.

Heat Transfer Methods

The transfer of heat from one location to another can take place in a variety of ways. The four most common methods are:

1) radiation
2) convection
3) conduction
4) evaporation

Radiation Heat Transfer. Whenever one object is warmer than another, heat energy will be transmitted across space (even in a vacuum) by radiation. The most common example is the sun—heat radiates across millions of miles of space and a tiny portion of this total solar energy is received on earth. Another example is the radiation of heat from the ground into the atmosphere on a cold clear night, which is sufficient to make the ground colder than the air near the ground, often resulting in ground, or radiation, fog. All that is required for radiation heat transfer is a difference in temperature between two surfaces.

Convection Heat Transfer. Heat can be removed from a warm object by the action or movement of air and water. For example, an exposed steam radiator may have a surface temperature of about 215 degrees. The room air next to the radiator becomes warm by contact, and rises because warm air is less dense, and therefore lighter, than cool air. The heat removed by this natural air movement is called *natural convection.* The same action can occur with a fluid, such as water. The bottom surface of a pan of water is heated by a burner, the heated metal surface warms the water in contact with it, the heated water rises in the pan because its density is less than that of cooler water, the warm water is replaced by cool water, and a continuous circulation of thermal energy is established.

The convection process can be assisted by means of fans, pumps, or other positive movement devices that increase the velocity of the air or water over hot or cool surfaces. This artificially induced circulation is called *forced convection.* A desk fan blowing over a hot

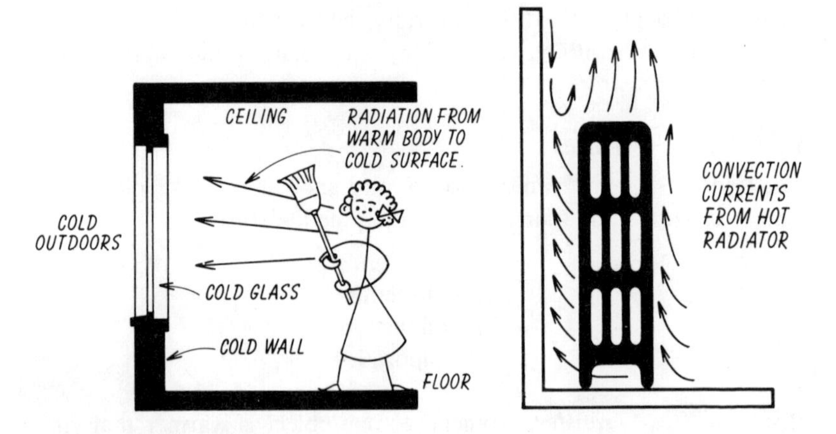

steam radiator will cause the heat output of the radiator to increase.

Conduction Heat Transfer. Heat can be transferred from one end of a solid object to the other by conduction. A common example is that of an iron poker thrust into a bed of glowing coals. Heat energy will move from the hot end toward the cooler end by conduction through the molecules of iron.

Another example is the sensation of warm or cold experienced through the soles of the feet when you are standing on a hot pavement or a sheet of ice. Heat is being transferred from the hot pavement through the soles of the shoe to the bottom of the feet; heat is also transmitted from the foot through the soles of the shoe to the sheet of ice. Heat *always* moves from a hot surface to a cooler surface.

Evaporation Heat Transfer. When water or some other liquid evaporates, a cooling action takes place because it requires heat to change a liquid into a gas. The best example of evaporation is the body bathed in perspiration on a warm day. The evaporation of the perspiration creates a cooling sensation because heat from the body is used to change the liquid into a vapor. Heat transfer by evaporation can also occur when tap water is spread over a basement floor—the water will evaporate into the basement air, the floor will be cooled, and the heat required for evaporation is removed from the basement air or from the floor. It takes about 8,800 Btu to evaporate 1 gallon of water.

2

Insulating Your Home

CONDUCTORS AND INSULATORS

Probably the single most important way of conserving energy in the home is to reduce the heat loss of the house itself by the use of insulation.

Though all materials conduct heat, some are much better conductors than others. For example, aluminum conducts heat more than 1,000 times better than wood. Materials that conduct heat slowly are classified as *heat insulators* and are often used to reduce the rate of heat loss from a house.

A desirable home insulating material should have low conductivity, be fire resistant, vermin and insect resistant, slow to settle and move when installed, and be unaffected by moisture or time.

Materials commonly used for home insulation include mineral wool, fiber glass, cellulose fibers, low-density particle boards, foamed plastics such as polystyrene and polyurethane, expanded mineral materials such as vermiculite or perlite, and reflective materials such as polished aluminum foil. These materials may come in the form of loose granules or tufts; folded into blankets or batts, either wrapped in foil or paper or unwrapped; flexible sheets; or rigid boards.

The Insulating Value: R. The insulating value of most insulating materials is printed on the wrapper with the code letter "R," which stands for resistance value. This value may be given for various thicknesses of materials or for the ways in which the material is installed. For example, insulating batts or blankets may have one "R" value for installation in sidewalls and a second value for use in ceilings for

COMMON FORMS OF INSULATION

exactly the same thickness, the difference being in the way the material is installed and the direction of heat flow.

Mineral Wool. Mineral wool and fiber glass are essentially the same type of material, which is formed by melting rock, slag, or glass, and extruding it into fine fibers, which are then stuck together with resins into tufts, blankets, or batts. It is completely fireproof, since it cannot burn, and melts only at very high temperatures. It has no food value or attraction for vermin or insects, and does not deteriorate with time, but may settle slightly. It does have one drawback. The fine fibers of this product may become embedded in the skin during handling and cause itching and a rash. Gloves and old clothing that can be discarded should be worn when handling these materials. Mineral wool has an "R" value of 3 per inch of thickness in sidewalls and somewhat larger in ceilings or roofs. Since mineral wool and fiber glass are the most common insulating materials, when we discuss inches of thickness here we will refer to mineral wool or its equivalent unless otherwise specified.

Cellulose Fiber. Cellulose fiber insulation is made from wood pulp, often recycled paper, which has been chemically treated to be moisture resistant and fire retardant. Though it will not burn by itself, it will burn if other combustible materials are involved. Extended exposure to water or moisture may cause some deterioration. It is available as loose fill, batts, and blankets, and it is easy and safe to handle. Some settling of loose fill may occur with time. It has about the same "R" value as mineral wool in all situations.

Particle Boards. Low-density particle boards are made from miscellaneous wood fibers and often appear as ½ inch or $^{25}/_{32}$ inch thick asphalt-coated, insulating sheathing used behind the siding of frame or masonry veneer construction. They are occasionally used to insulate crawl spaces or basement walls, or an uncoated version may be used behind interior-finish materials primarily because of its acoustic properties. It is water repellent, but extended exposure to water or moisture will cause deterioration, and it is subject to termites, molds, and mildew. It has an "R" value of 2.6 per inch.

Plastics. Foamed plastics, commonly polystyrene and polyurethane, are available as rigid boards of various thicknesses and sizes. They may be cut easily with a saw or knife and attached with nails or adhesive. Polystyrene is available in two forms, as cast closed-cell foam boards, or boards composed of expanded polystyrene beads stuck together by heat and pressure. Cast polystyrene foam acts as a vapor barrier, eliminating the need for a separate foil or plastic barrier. Though classified as self-extinguishing by some tests, both polystyrene and polyurethane will burn vigorously when other fuels are involved, giving off large quantities of toxic smoke. Polyurethane melts into a sticky gel, similar to napalm, as it burns. For these reasons foamed plastics should *never be used as exposed interior finish* in a home. Foamed plastics do not deteriorate with exposure to moisture, and have no attraction for insects or vermin. The insulating value of polyurethane foam does decrease slightly with extended age. Foamed plastics are very efficient insulators, with polystyrene foam having an "R" value of 4 and polyurethane of 6 per inch of thickness. They are often used where space is a factor, such as insulating refrigerators. In the home they are used to insulate the edges of floor slabs, the interior surfaces of masonry construction, and flat roofs.

Expanded minerals. Expanded mineral materials, such as vermiculite or perlite, are usually available in the form of loose granules used for fill in walls or ceiling spaces. Special waterproof versions of these materials may be used to fill the hollow cores or cavities of masonry construction. Some expanded minerals are also formed into boards for use in insulating flat roofs. These materials are completely fireproof, do not harbor insects or vermin, and do not deteriorate with age or moisture. Perlite board has an "R" value of 3 per inch of thickness in roof decks, and expanded mineral loose-fill insulation has an "R" value of slightly less than 2 per inch of thickness.

Reflective Insulation. This insulation, usually a metal foil or foil-surfaced material, differs from other types of insulation. An extremely smooth polished surface reflects heat back to the source, and also emits less heat on the cold side. Therefore, the number of reflecting surfaces and air spaces, not the thickness of the material, determines its insulating value. To be effective, each reflecting surface of the foil must be exposed to an air space at least ½ inch in thickness. Since heat is transmitted by convection as well as radiation, and air can move freely between reflective layers, reflective insulation is most efficient when used against downward heat flow, such as in a floor. In this application a sheet of reflective material, highly polished on both sides, is approximately equal to 2½ inches of fiber glass. Against horizontal heat flow, such as in a wall, a polished surface facing an air space has an "R" value of about 2. Reflective insulation is not efficient when used to prevent upward heat flow, because heat moves upward readily by convection. Since most reflective insulations are composed of metal foil, they are completely stable and fireproof, but the insulating value can be reduced if they are installed so that dust readily collects on the reflecting surface. Polished foils are often used on the back of dry wall or on the back of insulation batts as a vapor barrier, and if they face the appropriate size air space, they do contribute to insulating value as well. However, such a foil coating does not contribute to thermal resistance when the cavity is completely filled. It acts as a vapor barrier only.

INSULATING THE ATTIC

The first and usually the easiest place to check the insulation in your home is in the ceiling or attic. The ceiling or roof should have a total "R" value of at least 19. This can be achieved in normal construction by using the equivalent of 6 inches of mineral wool or cellulose fiber insulation.

If the house has an unoccupied attic or a space between the ceiling and a flat- or shed-type roof, insulate the attic floor.

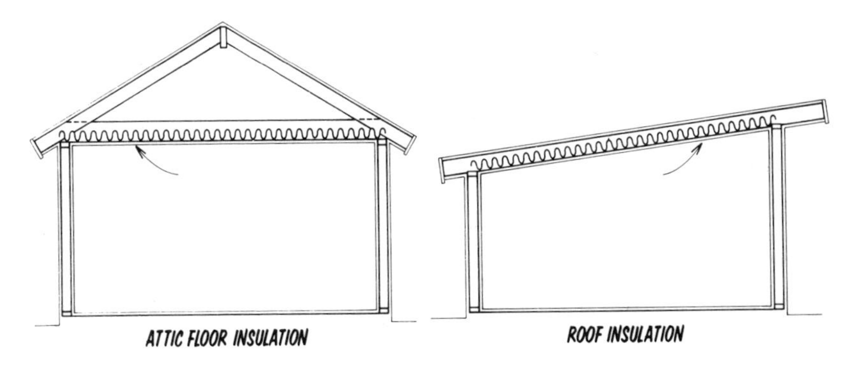

If the house has no space between the top floor ceiling and the roof, either because the attic is occupied or because there is a flat or shed roof with no separate ceiling construction, insulate the roof.

With an interior temperature of 75 degrees and an outdoor temperature of 0 degrees, an uninsulated ceiling of 1,000 square feet will lose 23,000 Btuh to a well-ventilated attic. Installing 6 inches of mineral wool or its equivalent in other insulations will reduce this heat loss to only 2,900 Btuh. Four persons sitting quietly give off about 1,800 Btuh. Additional insulation beyond the equivalent of 6 inches of mineral wool usually does not pay, because the heat loss through the ceiling has already become almost negligible. However, instances of 8-inch depths are often used to allow for some settling of the material. The feeling of comfort is greatly improved after installing 6 inches of insulation because the ceiling surface temperature, which would be 61 degrees without insulation, would be raised to 73.3 degrees with insulation.

If there is insulation already installed in the ceiling, but a less than adequate amount, additional insulation may easily be added, either in the form of another layer of blanket or batt-type insulation or by filling to the required depth with a loose-fill-type insulation. Both are easily installed by the homeowner, and if there is a floored attic, either batts or loose fill can be pulled or pushed between the board by removing one or two boards at various places across the attic. If no flooring is installed, the loose fill may be distributed with a

rake over the entire attic surface. Be sure to place a wide plank on the ceiling joists for a working platform—the ceiling will not ordinarily support the weight of a man.

VENTILATING THE ATTIC

All insulated attics should be ventilated; any water vapor that enters the attic space from the rooms below needs to be vented from the attic so that moisture does not condense on any cold surface in the winter. During the summer the heat that accumulates in the attic space must be vented or purged in order to prevent the attic from becoming overheated and eventually transmitting heat into the occupied space below.

Most houses have some form of louvers or slots along the lower edge of the roof, and either gable-end louvers, roof louvers, or a continuous ridge vent at the peak of the roof. Check the attic space to see whether natural ventilation can take place. If the attic space is entirely closed, some means for venting the area should be found. Make sure that the openings do not admit insects, birds, squirrels, leaves, rain, or snow.

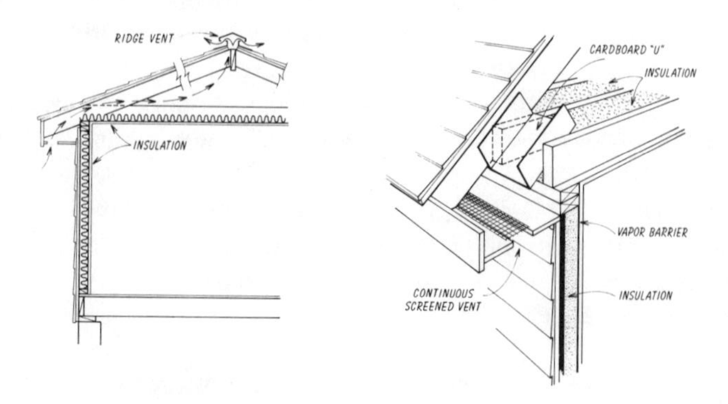

If the attic is adequately insulated with a 6-inch layer of mineral wool, or its equivalent, the use of an attic fan to blow outdoor air through the attic space in the summer will not be worthwhile. An attic fan may be worth considering only when the attic space cannot be conveniently entered for applying more insulation, and the usual means for naturally venting the attic space are difficult to install.

When adding insulation in the attic, care must be taken to be sure that the vents in the attic space are not plugged with insulation. A piece of cardboard, bent into a U shape and placed between each pair of rafters, will ensure that this ventilation space is kept open and will protect lightweight loose fills from being blown around by air movement.

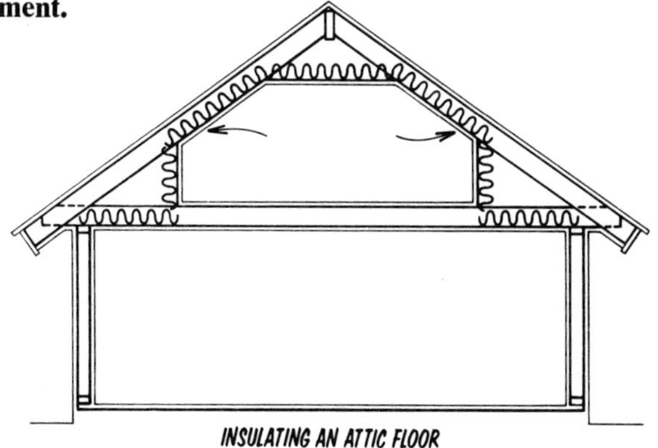

INSULATING AN ATTIC FLOOR

If there is an occupied area in the attic, its walls and ceiling must be insulated just as you would insulate the walls and ceiling of the house.

If there is no space between the top floor ceiling and the roof, either because the attic is fully occupied or because there is a flat or shed roof, it is very difficult to add insulation, and this may have to be postponed until the next time the building is reroofed. At that time the existing roof can be stripped off and additional board-type insulation can be installed prior to reroofing. You may wish to check with a competent roofer about the possibility of removing the gravel from an existing built-up roof, adding foam or plastic insulation above the roofing surface, and then reinstalling the gravel to hold the insulation in place. This type of "upside-down" roof construction has proven successful in some installations.

INSULATING THE WALL

The wall is the second largest surface area of the home that can lose a great deal of heat. For maximum saving, walls should be insulated to an "R" factor of at least 13.

In typical frame or masonry veneer construction, each 1,000 square

feet of uninsulated wall surface, excluding windows and doors, will lose 18,750 Btuh. If the normal frame or veneer wall, 3½ inches in thickness, is filled with mineral wool or blankets for a total "R" value of 15, this heat loss can be reduced to 4,725 Btuh. Also, the surface temperature of the wall will be raised from 62.3 to 71.8 degrees, which adds considerably to the comfort of the occupants.

Of course, the best time to insulate a wall is when the house is built. If, however, the wall cavity is completely empty in frame or masonry veneer construction, loose-fill-type insulation can be blown into place by a competent installer using specialized equipment. Usually this involves removing one or two rows of siding or brick at the top of the wall to permit holes to be cut into each stud space. A hose is inserted into these holes and the insulation is forced into the wall under air pressure. These holes are required at the top of each story of the house and below the windows. A competent installer will also check each studding space with a sounding weight to check for the existence of diagonal braces or fire stops—additional holes are required below such obstructions in order to completely fill all spaces in the wall.

In order to permit the escape of any water vapor that might enter the stud space from the living quarters, the installer will probably provide small air vents in each of these spaces.

If the wall cavity is partially filled with batt or blanket insulation, it is not feasible to add additional insulation because the material already present keeps the new fill from flowing into place. It is seldom worth the expense of removing either the interior or the exterior finish to add insulation to a partially insulated wall, since new insulation will not greatly increase the effectiveness of the existing insulation in most cases.

In the case of a solid masonry house, if the masonry is concrete block, one of the expanded mineral fill-type insulations can be poured into the cavitities of the block to increase the insulating value of the wall. Though this again is best done during construction, this insulating material can be added if the top plate on the wall does not completely cover all the block cores.

If the construction is of the cavity wall type, masonry fill insulation can be poured into the cavity from the top, although any

excess mortar left within the cavity by the masons may prevent the fill from going all the way to the bottom of the wall.

INSULATION HELPS TO KEEP BASEMENT WALLS WARM; THUS, WATER VAPOR DOES NOT CONDENSE ON THEM READILY.

The best way to insulate solid masonry construction is by applying board-type insulation to the inside surface of the wall, beneath the plaster or dry wall. These boards may be applied between furring strips attached to the masonry or in some cases applied directly to the masonry with an adhesive, with plaster applied directly over the insulation. Applying a 1 inch thick foamed plastic board between furring strips on a solid brick wall will at least double the insulating value of the wall (R=10). Special trim around doors and windows will be necessary to cover the increased wall thickness. Because the thickness is critical, foamed plastic is recommended for this application because of its greater insulating value per inch of thickness.

INSULATING THE BASEMENT

An often overlooked area of heat loss is through the floor system, either slab, crawl space, or basement.

When the basement is used only for storage, the floor above the basement should be insulated and the basement maintained at the lowest practical temperature. The band joist should be insulated with a 2-inch layer of batt insulation and the same material placed along

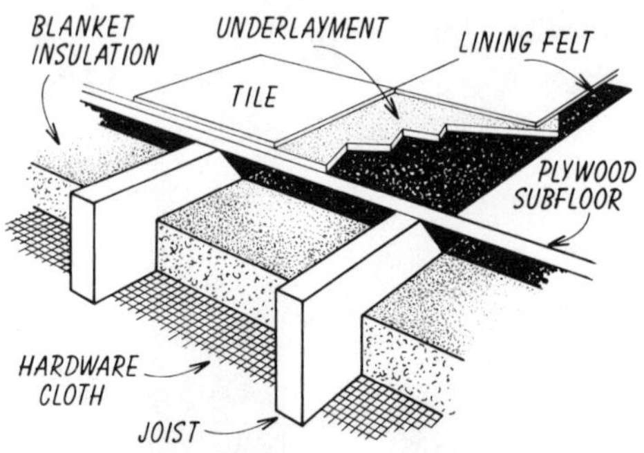

FLEXIBLE INSULATION, INSTALLED IN THE SPACE BETWEEN THE JOISTS UNDERNEATH THE FLOOR, SHOULD BE SUPPORTED BY HARDWARE CLOTH, WIRE NETTING, OR A SHEET MATERIAL FASTENED TO THE BOTTOM EDGE OF THE JOISTS.

the bottom of the floor joists. The batt or blankets can be held in place by a supporting wire mesh. As an alternative, a layer of double-polished reflective insulation may be stapled to the bottom of the floor joists.

When the basement is used for living quarters, and the air temperature is maintained at the same level as in the rest of the house, more extensive insulation is recommended. Obviously, the windows in the basement should be provided with double glazing. The band joist should be insulated with 2 inch thick batt-type or board-type insulation. Basement walls should have at least 2 inches of insulation installed from the top of the wall to a point at least 2 feet below the outside ground level. For example, an uninsulated basement 30 by 50 feet will lose 21,520 Btuh when there is a 75-degree indoor-

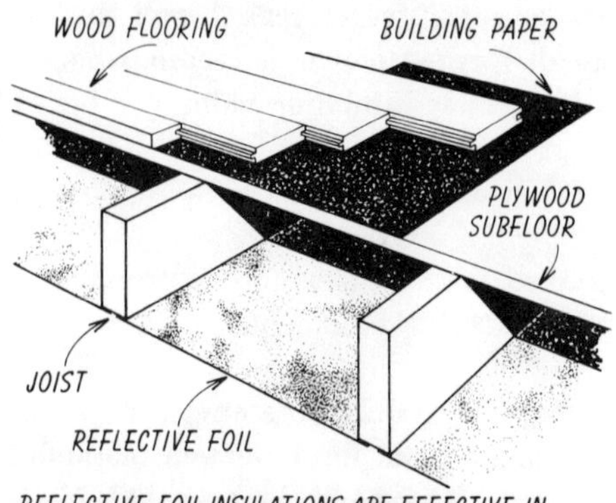

REFLECTIVE FOIL INSULATIONS ARE EFFECTIVE IN INSULATING FLOORS ABOVE OPEN CRAWL SPACES AGAINST DOWNWARD HEAT LOSS.

outdoor temperature difference. However, when properly insulated the loss can be cut to 11,400 Btuh.

INSULATING THE CRAWL SPACE

If the house is built over a crawl space, two options are open to the homeowner.

For those crawl spaces that are not tightly enclosed with walls, the only recourse is to insulate the band joist and then place insulation along the bottom of the floor joists (as in the case of the basement used only for storage). This procedure will not only reduce the heat loss through the floor, but will provide a warmer floor.

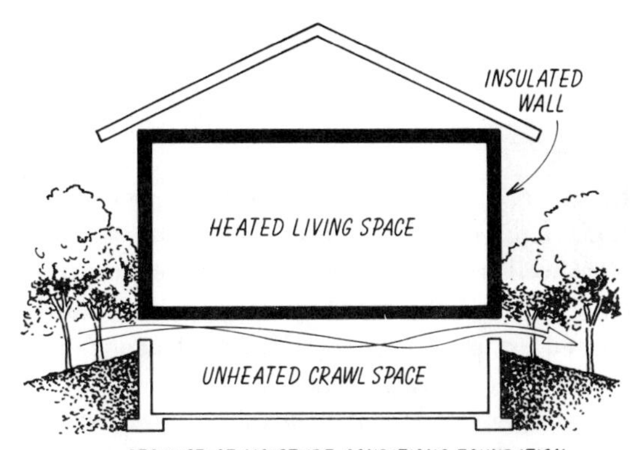

BECAUSE OF MOISTURE CONDITIONS, FOUNDATION VENTS IN SOME CRAWL SPACES MUST BE KEPT OPEN THE YEAR ROUND.

A mobile home without skirt boards is a modification of the above case and the same recommendation applies.

Still another modification of the above is the case of the bedroom placed over an attached garage. The floor of the bedroom is over an unheated space that is frequently left open. The floor of the bedroom should have ample insulation (4 to 6 inches deep) between the floor joists, and fire protection (such as ⅝-inch UL listed gypsum wallboard) for the garage ceiling. A continuous sheet of plastic film, placed over the floor joists before the flooring is applied, will not only serve as a vapor barrier but also as a means of preventing car engine fumes from entering the bedroom.

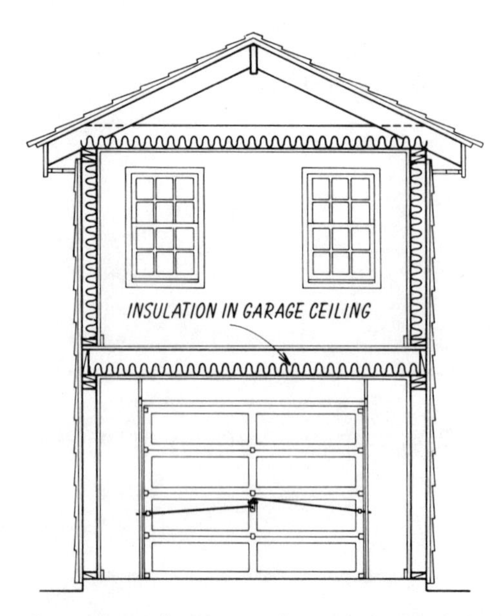

If the climate is moderately severe, the crawl space should be tightly enclosed within a masonry wall. Since there are water pipes in the crawl space, as well as heating ducts, it would be preferable to insulate the band joist and the crawl space wall. The crawl space vents should be closed to prevent excessive air leakage into the crawl space. The walls of the crawl space should then be insulated to a depth of at least 2 feet below the outside ground level, if practical, with 2 inches of board-type or batt-type insulation.

Since the crawl space vents are to be closed (in winter), the installation of a 6-mil polyethylene vapor barrier over the entire ground surface is an essential part of the solution. This will tend to eliminate excessive moisture in the rest of the house, a condition that is often encountered in houses with crawl spaces. The polyethylene vapor barrier should just be unrolled over the floor surface of the crawl space, with the joints in the polyethylene sheets overlapped by at least 6 inches, and the vapor barrier extending up the wall over the insulation. If foamed plastic board-type insulation is used, it is not necessary that the vapor barrier extend over the insulation.

In cases where poor water drainage is a problem, closing the crawl space vents may not be practical. In those cases where the house is

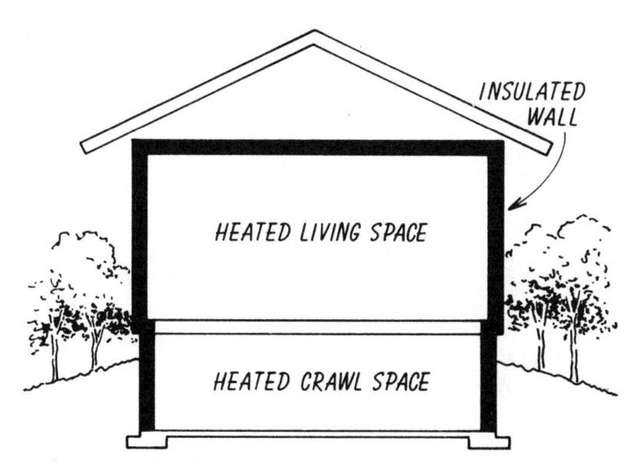

A CLOSED CRAWL SPACE IS ONE THAT CAN BE HEATED. USE OF A GROUND COVER IS ESSENTIAL FOR THIS TYPE OF CRAWL SPACE.

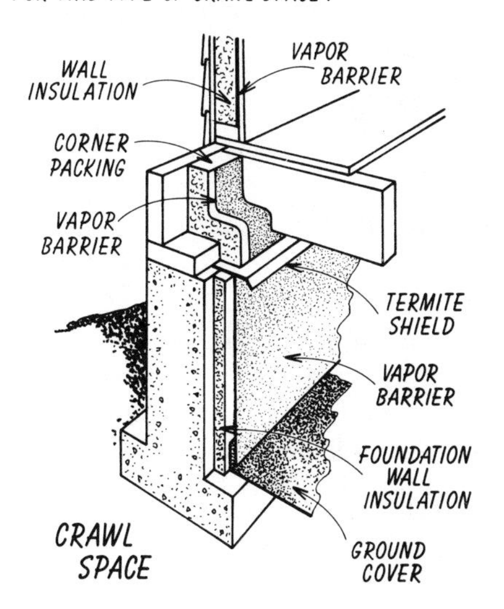

located in a hollow, and heavy rains wash into the crawl space, the first order of business is to prevent the water entry by diversion or drainage. If the crawl space is waterlogged part of the winter, it may be necessary to keep the vents open to dry out the area as much as possible. The house should then be treated as one with an open crawl space.

INSULATING THE CONCRETE SLAB FLOOR

If the house is built on a concrete slab on the ground, it is most important for both heat retention and comfort that the *edge* of the slab be insulated. This is especially true for slabs that have embedded heating ducts at the perimeter of the slab, and to a lesser extent for hot-water baseboard radiators at the outer edge.

This edge insulation is best accomplished at the time the slab is poured. Foamed plastic insulation can be installed either vertically along the inside surface of the foundation and extending through the slab or as a short piece extending up past the edge of the slab and turned under the slab by at least 24 inches.

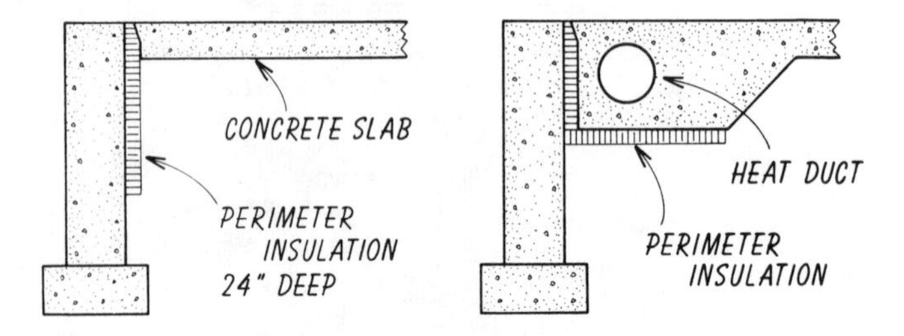

An existing house with an uninsulated slab presents a special problem. The insulation can be applied to the outside of the foundation wall, again extending at least 2 feet below ground. Obviously this requires digging dirt and exposing the construction of the foundation. The insulation that remains exposed to the weather, after the dirt is replaced, should be protected by a weatherproof material such as cement-asbestos board. In addition, a flashing should extend under the bottom row of siding and over the upper edge of the insulation and finish material.

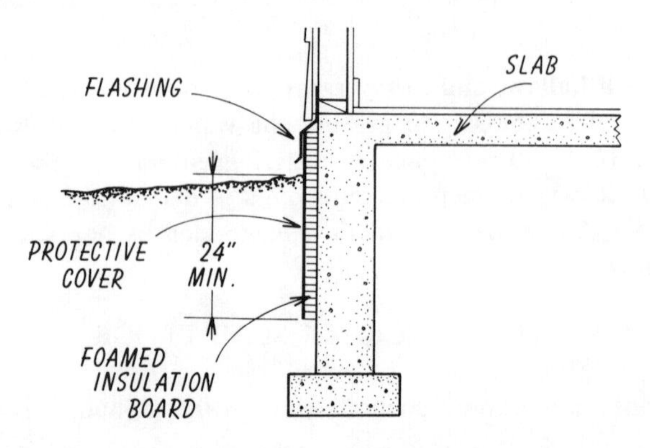

This same system may be used to insulate an uninsulated breezeway or garage slab, which is being converted into living space. The difference in heat loss between a properly insulated slab and an uninsulated slab floor is 30 to 40 percent, depending upon the location of the heat ducts. Note that a concrete floor slab that forms a basement floor (several feet below ground surface) need not be specially insulated, because the heat loss is small and any insulation placed below the slab would be ineffective.

INSULATION FOR SUMMER COOLING

Though all the above examples have been based upon saving energy by insulating a house for winter heating, savings are also possible where summer cooling is employed, even though the temperature differential between indoor and outdoor air is usually much lower during the cooling season.

One of the greatest benefits of adequate insulation between a ceiling and a hot attic is a greatly reduced ceiling surface temperature. The uninsulated ceiling can reach surface temperatures in late afternoon that provide a "broiler" effect. In fact, the entire ceiling surface becomes a heat panel that radiates heat to the occupants below.

When adequate ceiling insulation is provided, another measurable effect is the shift of the maximum heat gain into the house from about 3:00 P.M. to as late as 6:00 or 7:00 P.M. The heat gain through the ceiling, therefore, is delayed several hours past the peak outdoor temperature. This means that a smaller cooling unit can be installed and its operation is not going to aggravate the "peak" power loads currently experienced by all electrical utilities.

Our largest potential energy savings can be made in those southern areas where insulation has been largely ignored to date. Ceiling and wall insulation will permit the installation of smaller cooling units, which will result in lower operating costs. In these warmer climates proper insulation is of benefit primarily in summer weather, but many fringe benefits will be experienced during the heating season as well.

WEATHER STRIPPING AND CAULKING

Infiltration of outside air is another major drain on energy in the home. During the winter the incoming air, which may already have a low moisture content, must be heated, and this tends to further reduce the relative humidity in the house, creating a dry air problem. In the summer the incoming air must be cooled and usually dehumidified.

Weather stripping and caulking play an important part in reducing infiltration. Wind can cause a buildup of pressure on a portion of the house, forcing air through even the smallest cracks. If a window is not properly weather-stripped, it may even be possible to see the curtains move when there is a heavy wind. (When the wind velocity is doubled, the air leakage increases about four times.) Since nearly all building materials expand and contract with changes in weather, enough clearance must be allowed for this when constructing the house. This expansion space must then be closed with some form of weather stripping or caulking. If a door must be moved regularly, weather stripping is usually used to seal the crack. If the crack is between two fixed portions of the house, then caulking is forced into the joint to close it.

Any crack around a window or door that is loose enough for a dollar bill to be inserted and pulled out needs weather stripping. Weather stripping of windows and doors that are not equipped with storm windows and doors may save as much as 3 percent of the total energy used in the house during the heating season.

Heating engineers have determined the following air leakages for various types of seals:

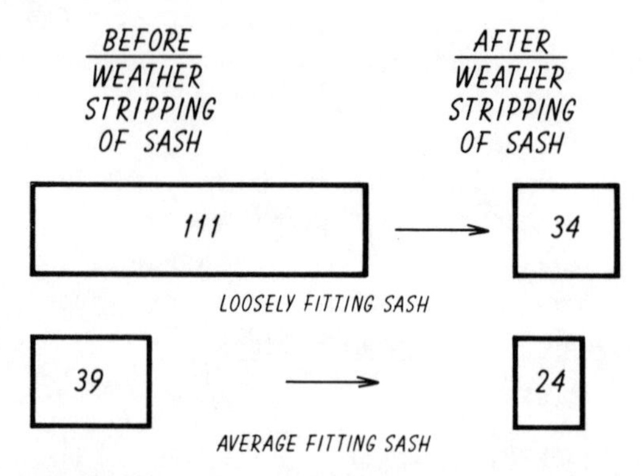

If storm windows are of the type that are completely separate from the prime window, weather stripping is less important because there is enough of a double seal against outside air—once by the storm window and once by the prime window.

WEATHER STRIPPING

There are essentially two types of weather stripping: one which depends upon a mechanical interlocking of two parts, and the other which depends upon the compression of some resilient material between one or two moving surfaces.

Mechanical weather strips are most often used on doors. They consist of two parts, each being hook-shaped, one of which is installed on the doorframe while the other is notched into the door edge. The second type mounts on the door stop, with the interlocking piece mounted on the door face. As the door closes, these hook-shaped pieces interlock with a spring action, preventing air passage. A similar arrangement is often used on door bottoms, where one hook is part of a metal threshold and the other attached to the bottom of the door.

Another type of weather strip that uses both mechanical and compression action is often used on the bottom of doors that must open over thick carpeting. In this case a compression-type weather strip is fastened to the bottom of the door at a point high enough to clear the rug, and it is lowered by mechanical action as the door approaches the threshold, forcing the felt, vinyl, or rubber strip down onto the threshold as the door reaches the closed position.

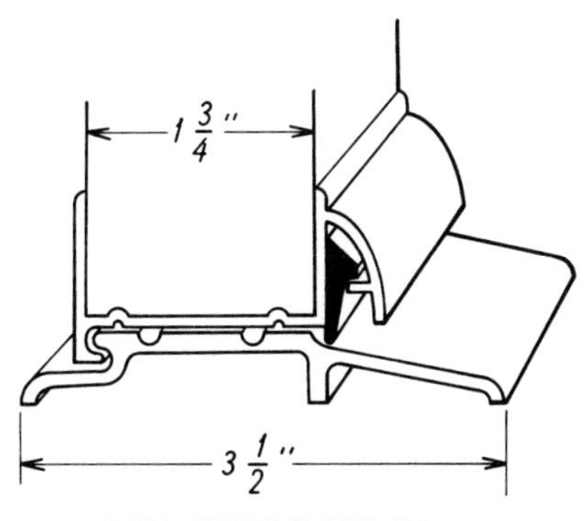

INTERLOCKING THRESHOLD

COMPRESSION WEATHER STRIPS

The longest lasting weather strip is spring bronze. This comes in various shapes to fit almost every type of sliding or closing door but it is the most expensive and difficult material to install. Wind may also cause the strip to "hum" under certain conditions.

Where two surfaces must remain in close contact but still slide, the most popular type of weatherstrip is a wool or synthetic pile. This is the type used by most manufacturers for factory weather stripping of windows. These synthetic piles wear well and remain springy for many years.

Probably the most popular type of weather strip used by do-it-yourselfers is foam rubber or foam plastic, either adhesive-backed or attached to a wooden strip. Foam rubber is somewhat more durable than plastic, but takes more force to compress and does not compress into as small a space as does foam plastic. If there are considerable

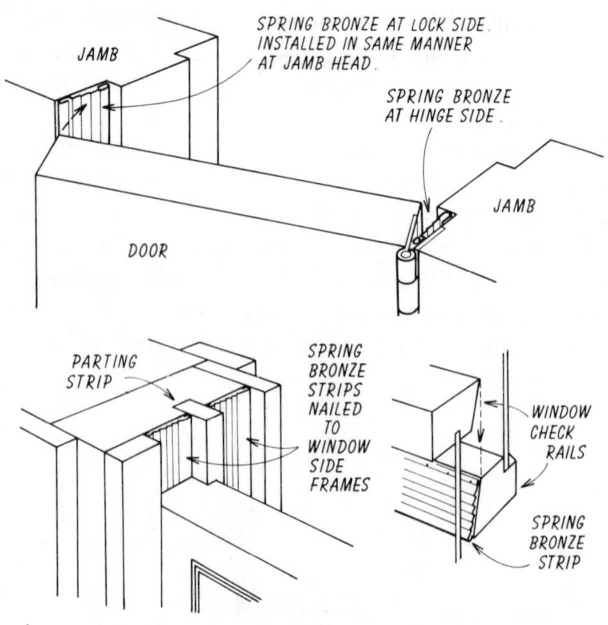

variations in the space to be closed, foam plastic may be superior. A variation on this type of weather strip is vinyl or rubber tubing, sometimes foam-filled, which is compressed between two meeting surfaces.

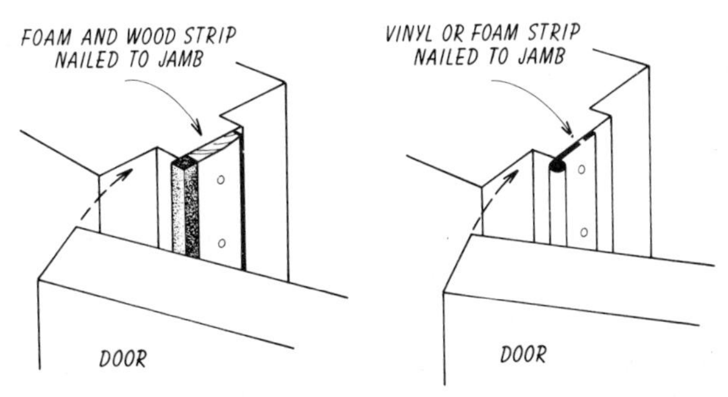

The cheapest variety of compression weather strips is felt, but it has the least amount of resilience to accommodate variations in the gap to be filled. Also felt weather stripping has little resistance to abrasion and it therefore used where two materials come together in compression rather than in a sliding fit.

Molded rubber or plastic strips are often used on door bottoms and thresholds. In the case of the threshold it is restrained at the edges with the center bulging up to meet the door. In the case of vertical operation, such as garage doors, the molded strip is nailed to the bottom of the garage door with the projecting edges coming down and conforming to the garage floor.

One very efficient weather strip for specialized application is the vinyl strip with enclosed magnets that snap the strip against a metal door. This is the type commonly used on refrigerator doors and is often standard equipment on metal-clad residential doors.

A fringe benefit that occurs with the application of tight-fitting weather strips on windows and doors is the reduction of outdoor dust

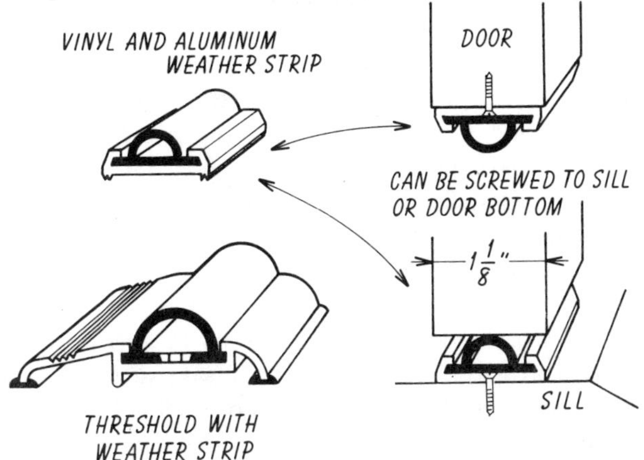

and dirt that is blown into the house. Another fringe benefit is that street and traffic noise are reduced to some extent.

CAULKING

If there is a gap where two materials meet, such as siding and window trim, or corner trim, any cracks must be caulked. This is primarily to keep water from penetrating the structure of the house, but it also helps to prevent air infiltration. Caulking works like chewing gum in any crack or opening to prevent wind and water penetration. The ideal caulking will adhere to both sides of the opening to be closed and remain resilient to permit movement between the two materials without cracking. Inexpensive caulking may be able to withstand only minor expansion or compression before it fails and cracks. Good caulking may withstand a great deal of expansion or contraction before failure.

Caulking should also be used to close any cracks in masonry construction, either in the siding or foundation. In brick masonry construction there are often weep holes left in the vertical mortar joints in the bottom row of bricks to permit any moisture that penetrates the wall to drain out.

Do not caulk or block the weep holes in any way. If there seems to be an air infiltration problem, a length of fiber-glass rope can be stuffed into the weep hole, which will effectively block air circulation, but will withdraw any water from behind the brickwork like a wick. If the prime window in the house is relatively loose, it may be simpler to caulk the storm sash into place rather than to weather-strip the prime one. However, in no case should the bottom of a combination storm-screen sash be caulked. When the storm panel is raised and the screen is in place, a blowing rain will penetrate through the screen, and if the bottom of the sash is caulked the water will accumulate on the windowsill and run into the house. Most such storm-screen combinations have holes along the bottom edge to allow for drainage. They should never be plugged.

It should also be noted that caulking the storm sash into place may cause problems with condensation on the inside surface of the storm window if the storm sash is tighter than the prime one. The cure in this case is to make the prime sash tighter so that water vapor does not leak past the prime sash.

If windows will not be opened during the heating season, a special type of caulking, in the form of a coiled string, is available for one-season use, and it may easily be stripped off in the spring. This caulking cord is relatively inexpensive and easy to handle. It is applied with the fingers and cut with scissors.

An alternative to this type of caulking is the use of masking tape, either on the inside or outside of the window, for those windows that will not be operated during the heating season. Fabric-backed heating duct tape may be more suitable than masking tape, since it has less tendency to leave an adhesive on the window trim when it is removed.

There are several common types of caulking materials. The most common and least expensive is white lead-based caulking. This usually lasts only one or two years before cracking, and can withstand very little elongation or compression before failure. It is available in both knife grade and in tubes for use in a caulking gun.

Latex and Butyl caulks are made with a synthetic rubber base, and are able to withstand a moderate amount of elongation or compression without failure. Their estimated life is at least five years, and they can be painted with both oil-base or water-base paints. They are also somewhat softer and easier to apply from a caulking gun. The cost is usually about four times that of white lead-base caulks.

The best of the caulking compounds are of the hypalon, polysulfide, or silicone types. They are usually applied by professionals, who know exactly what type to use for a specific situation. They are long-lasting, capable of withstanding a great deal of elongation or compression, but they are expensive, often fifteen times as much as white-lead caulk. Silicone caulking is available to the homeowner as bathtub sealer, and in this form it is often used by do-it-yourselfers. These caulks usually cannot be painted with any conventional paints, but they are available in various colors.

Caulking compounds are available in knife grade, which is applied by putty knife or special applicator. Perhaps the most convenient way for the homeowner to purchase caulking is in the form of a paper cartridge with a plastic nozzle, which is inserted into a hand-operated caulking gun. These guns are relatively inexpensive, usually only $1 to $2. The cartridges are inserted into the gun, the tip is cut to the proper shape to fill the type of crack involved, the seal inside the tip is broken, and the trigger-type handle is worked to produce a slow stream of caulking from the tip.

The caulking gun should always be moved in the direction in which it is pointed, pushing a small bubble of caulking material

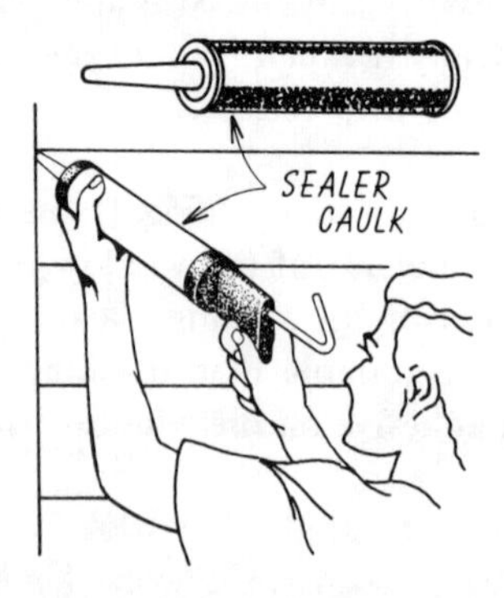

ahead of the tip and allowing the tip to force this caulking material into the crack. If appearance is of prime importance, the caulking can be smoothed with a wet finger or tool immediately after application. The surface of the caulking will usually harden within a half hour to the point that insects or dust will not stick to it. This is one area of home maintenance that is relatively easy for the inexperienced homeowner to perform himself.

DECORATIVE INSULATION

Decorative insulations are materials applied to the interior of a house for the purpose of conserving heat as well as making the home more attractive. The hanging of draperies can save a great deal of energy. Heavy draperies drawn fully across a window can reduce the heat loss in the winter by 25 percent. If, however, the draperies also cover a heat outlet, such as a convector, diffuser, or radiator, they could increase the heat loss by channeling warm air against the window. To retain heat and save energy, the warm air should be introduced on the room side of the drapery, not on the window side.

Some draperies are available with an insulated lining, either of the foam or reflective type, and these are somewhat more effective than conventional drapes.

A tightly fitting window shade, venetian blind, or inside shutter can also reduce the heat loss through a window by about 25 percent.

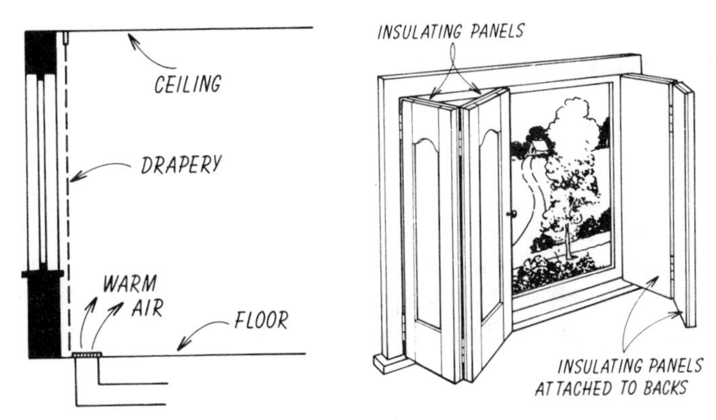

If inside shutters are used to cover windows during cold weather, the portion of the shutter that faces the window can be lined with a thin decorative foam plastic board, which may double the insulating value of a window equipped with insulating glass or a storm window. This paper-covered foam-core board is available at art supply and advertising display outlets. The board can be covered with fabric or be papered, painted, and so on for a pleasing exterior appearance.

Closing draperies, shades, blinds, or shutters does lower the temperature of the glass and it may cause condensation or frost formation if the humidity in the house is high. In fact, the presence of condensation is a good measure of the effectiveness of this method of insulation.

Decorative window coverings should, of course, be opened when the sun strikes the window in order to obtain the maximum solar heat gain. Draperies, shutters, blinds, and shades should be closed at night and during periods of overcast sky or high winds.

In the summer this treatment of decorative window insulation should be reversed. The window should be covered when the sun strikes it, and the draperies should be opened at night to allow as much heat as possible to radiate into the cooler night air.

With heavy draperies you are no longer exposed to cold window surfaces. Heat radiation from the body to the window surface is reduced.

With heavy full-length draperies, cold-air currents that seep through

the window are slowed by the drapery near the floor. Cold-air drafts are less noticeable across the ankles when draperies are closed.

Though many people believe that wood paneling adds significantly to the insulating value of a wall, if the paneling is substituted for conventional dry-wall products, it has no greater insulating value than the product replaced. If wood paneling is installed over an existing wall finish, it will help to a small extent. Quarter-inch wood paneling has an "R" value of approximately 0.25. This is less than 2 percent of the desirable "R" value of a well-insulated wall.

3

Storm Windows and Storm Doors

STORM WINDOWS (STORM SASH OR PANELS)

Glass surfaces are a significant factor in the energy consumption of a house in both winter and summer. Glass is a very poor insulator and is a major source of heat loss in the winter. If a well-insulated house has as little as 10 percent of the wall area in glass (the minimum permitted by most building codes), as much as 25 percent of the total heat loss may be through the glass. In the summer, glass is not only a poor insulator, but also admits the radiant energy of the sun, a large part of which is converted to heat. Storm windows or insulating glass reduce the heat loss through windows by more than half.

Storm windows are more efficient than insulating (factory-sealed double-pane) glass, although the latter is often preferred because there is nothing to put on in the fall and take off in the spring, and there are two less glass surfaces to be cleaned. The storm window is superior because it provides a separate seal of the cracks around the window, and also because the air space between the two layers of glass is greater, and the air space is an insulator. If your windows are not factory equipped with insulating glass, it will usually be necessary to replace the entire sash in order to install this type of glass. Therefore, the separate storm sash is the most economical solution.

Types of storm window insulation are best compared by referring to their "R" values.

When these values are compared with the recommended "R" 15 value for an insulated wall, it is easy to see why windows are such a great source of heat loss.

As a matter of fact, triple glazing (insulating glass plus a separate

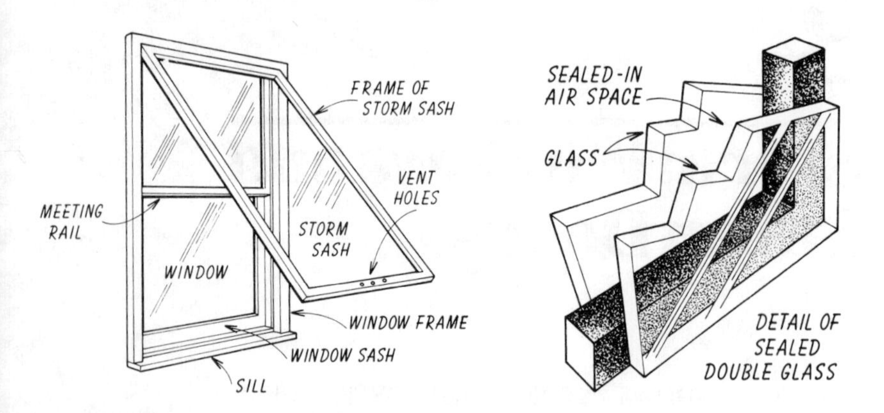

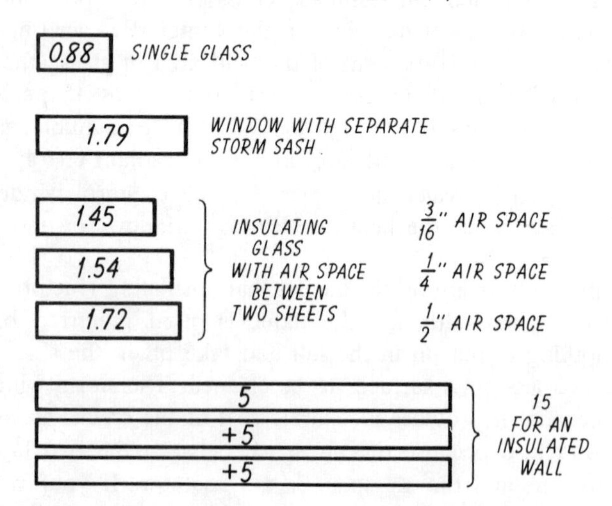

storm sash, or a new factory-sealed triple unit) can often be justified in areas with 6,000 or more seasonal degree days in houses that are also centrally cooled.

Triple glazing is also a more effective barrier in keeping outside noise from being heard in the house. It is sometimes used for this reason alone in houses near airports, for example.

In addition to reducing heat loss, the occupants of a room where the windows are equipped with storm windows or insulating glass will be much more comfortable. If the outside temperature is 0 de-

R VALUES FOR TRIPLE GLAZING

2.13	$\frac{1}{4}$" AIR SPACE FOR INSULATED GLASS
2.78	$\frac{1}{2}$" AIR SPACE

grees, the inside temperature of the glass surface of a single-glazed window is only 17 degrees—somewhat colder than the walls of an igloo. The inside surface temperature of insulating glass or a window equipped with a storm window will be 47 degrees. Triple glazing will provide an interior surface temperature of 57 degrees.

Also, the probability of moisture condensing on windows depends on the temperature of the glass and the amount of water vapor in the air within the house. When the outdoor air is 0 degrees, moisture condensation will occur on the glass when the relative humidity inside the house is only 12 percent.

With insulating glass, condensation will not occur until the relative humidity is 30 percent.

A window equipped with a storm window will not condense water until the relative humidity is 37 percent.

If moisture condensation occurs on a window equipped with a storm window, the humidity in the house is too high, and moisture will condense within the walls. In such an event the humidity should be reduced if condensation persists for more than 30 minutes.

Double-Hung Windows. The conventional double-hung window is usually equipped with a separate wood- or metal-frame storm sash attached at the outside of the window casing. Metal-frame storm sash are lighter in weight and easier to handle, and often are sold as combination storm-screen units; half the window is equipped with a screen and the storm panel for that half is stored as a second layer of glass in front of the storm panel during the summer. These units are convenient in that they do not have to be put up and taken down twice a year, and the individual storm panels are removable for washing. However, metal-frame storm sash are somewhat less efficient than wood-frame units because of the much higher heat conductivity of the metal. Aluminum-frame combination storm and screen units may be as much as 20 percent less efficient than a conventional wood-frame storm sash.

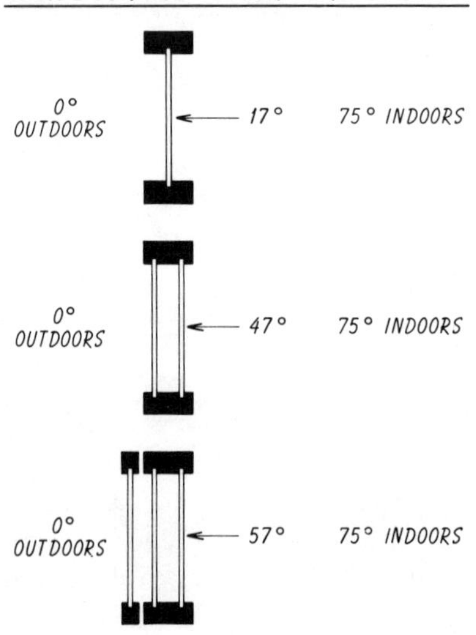

If the double-hung window is aluminum-framed and fits in a metal casing, a metal storm window should be installed in such a way that there is no continuous metal extending from outside the storm panel all the way to the inside of the house. There should be a thermal break of wood or plastic between the storm window and the metal prime window. One of the most significant benefits of storm windows

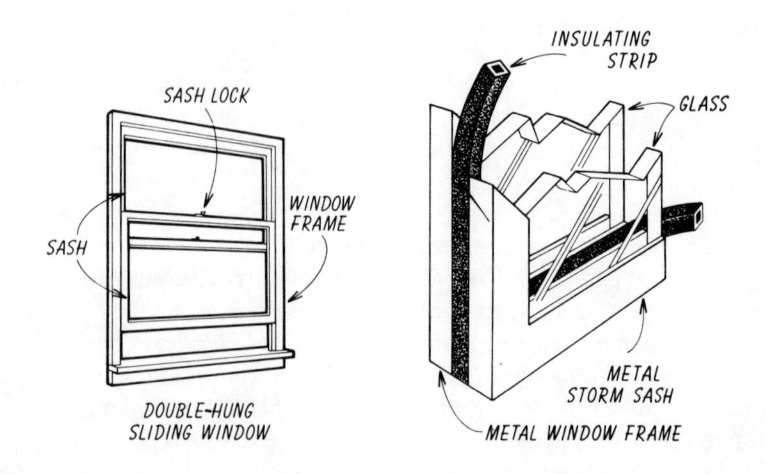

on metal-framed prime windows is a reduction in the amount of condensation that may occur on a metal window frame. Though this is not an energy-related problem, it is a definite housekeeping problem.

If moisture condensation occurs on the storm window, but not on the prime window, this is an indication that the prime window fits more loosely than the storm window. The solution to this problem is to weather-strip the prime window, or to ventilate the space between the windows by drilling several ¼-inch holes through the storm sash.

Horizontal Sliding Windows. The sash in this type of window slide horizontally. Usually there are two movable sash; sometimes one is fixed. These windows may be equipped with insulating glass, a conventional storm sash attached to the outside window frame, or a storm panel that attaches directly to the window sash. A storm panel is a pane of glass set in a narrow frame that can be clipped either to the outside or inside of a window sash. Panels can be used only on sash that have special hardware or a groove into which the panels can be fitted. Unlike storm sash, storm panels move with the operating sash and do not interfere with the operation of the window or the ventilation. If storm panels are used, the window must be weather-stripped since there is no secondary seal as provided by the storm window. Though it is more convenient to install a storm panel inside the window, it is preferable on the outside, since condensation may occur if the inner panel is not tighter than the outer panel. If the storm panel must be installed on the inside, it should be tightly weather-stripped or gasketed.

Casement Windows. A casement window consists of a sash hinged at the side to swing outward. Usually two or more sash, separated by a vertical divider or mullion, are used in one frame. Because of the way this window operates, the storm window must be located on the inside. Storm panels attached directly to the sash are often used. If an inside storm window is used, no ventilation is possible during the heating season.

Awning Windows. Awning windows are hinged at the top. They are manufactured as a single unit or as several sash stacked together in one frame. When open, the sash project at an angle like awnings.

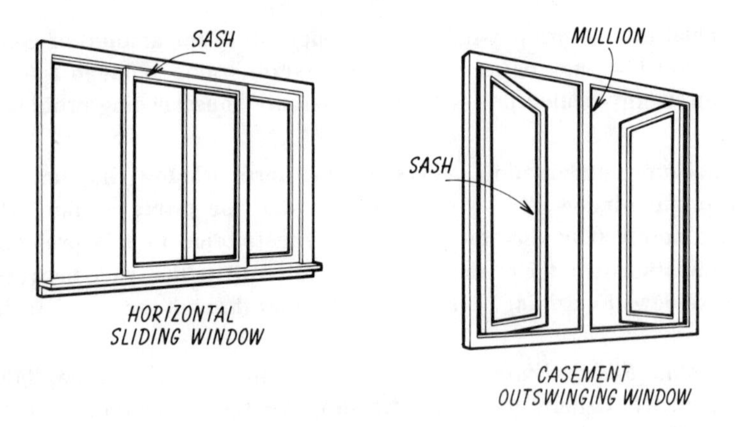

Again, inside storm sash or storm panels must be used, with the same limitations as described for casement windows.

In-Swinging Windows. Bottom-hinged or hopper windows and top-hinged windows often used in basements can be equipped with either storm sash on the outside of the window or storm panels attached to the prime window sash. No special problems are involved with either storm sash or storm panels.

Jalousie Windows. A jalousie window consists of a series of small horizontal glass slats, which are held by an end frame, pivot in unison

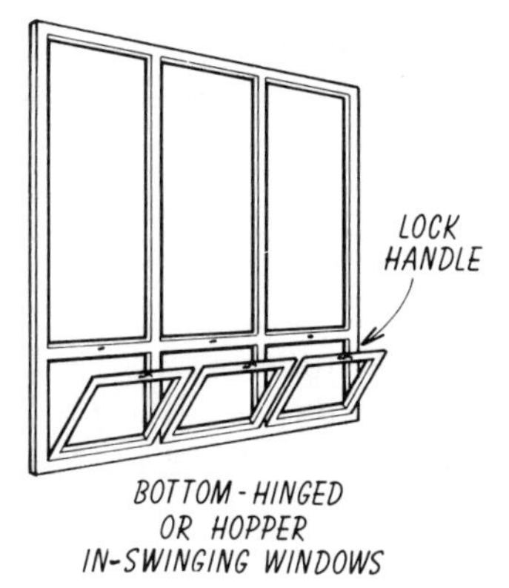

BOTTOM-HINGED
OR HOPPER
IN-SWINGING WINDOWS

like a venetian blind, and open outward. Jalousie windows are never satisfactory as prime windows for an area to be heated. Storm panels can be installed on the inside, but even with storm panels the large number of openings between slats results in cold air leakage in winter.

If you are a tenant or temporary resident, and do not wish to spend money on permanent sash or storm panels, temporary storm windows can be made using sheet plastic. The cheapest form of plastic storm windows utilizes polyethylene, which is available from any lumberyard. Storm-window kits, complete with polyethylene and tack strips, are often available from hardware stores. The polyethylene film may be installed either on the outside of the window, using tack strips, or across the inside of the window using double-sided tape. Polyethylene has the disadvantage of not being completely transparent, thus inter-

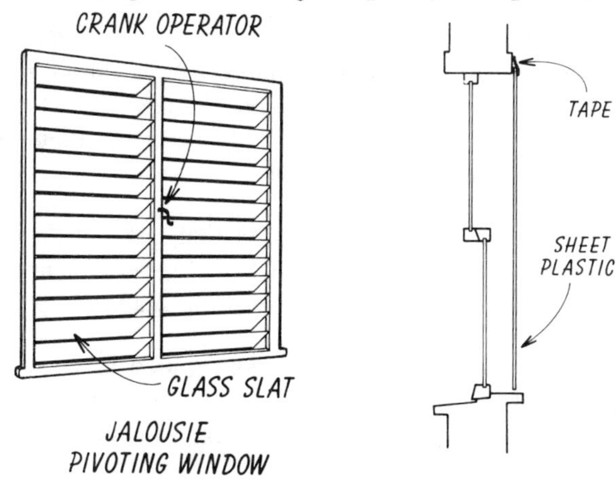

JALOUSIE
PIVOTING WINDOW

fering with the view through the window. It also deteriorates with exposure to ultraviolet light, and the useful life of polyethylene storm windows is no more than one season.

Vinyl film is available, which is somewhat more expensive than polyethylene, but it is clear. It does distort light to some small degree, much like the window glass made at the turn of the century. It is stronger than polyethylene, with an estimated useful life of about three years when attached to the outside of the window and five years when attached to the inside. It is installed in the same way as polyethylene.

A more drastic but more effective solution to glass areas that are not needed for light or view in the winter is to cover the entire area with a solid panel. This panel may be an inch of polystyrene foam insulation covered with a decorative surface, and held in place with clips, double-sided tape, or small nails. This might be particularly applicable for covering one or both panels of a sliding glass door, which leaks a tremendous amount of heat and which is seldom used in the winter. A panel could be applied to either the inside or outside of the window or glass door, depending upon the type of frame and appearance desired.

Though plastic foams are technically classified as self-extinguishing, they will burn vigorously when other flammable materials are involved and give off large amounts of toxic gases. For this reason exterior installation is preferred.

In the winter be sure to remove all screens from windows that get the winter sun. A considerable heat gain can be achieved by allowing even winter sunlight into the room, and a screen can block out approximately 50 percent of the available sunlight. Painted screens further reduce the amount of sunlight.

A special storm-screen unit is available that has a rollaway plastic screen attached to the bottom of the lower movable storm sash, so that the screen is out of sight when the storm sash is in place.

Conversely, in the summer any windows exposed to the sun may be equipped with screens or metal louvers which can significantly reduce heat gain. Louvered-type screens, however, will interfere somewhat with vision.

STORM DOORS

Outside doors exhibit considerable air leakage around the frame and the edges of the door. Heating engineers consider the air leakage around the door and frame as one of the most vulnerable spots in a house—an exterior door is considered to be as leaky as a loosely fitted window, on the order of 111 cubic feet of air leakage per hour for each foot of seam.

Caulk completely around the outside of the doorframe, and weather-strip all exterior doors.

The next step in reducing heat loss through doors is to install a storm door. A wood storm door with about a 50 percent glass area can reduce the heat loss through the door by 45 to 50 percent. A metal storm door, regardless of the percentage of glass, will reduce the heat loss by only 30 to 35 percent. (This assumes that the door is tight-fitting and not warped.)

The type of storm door that has a wood frame with about an 80 percent glass area, to better display unusual or decorative prime doors, has an insulating value similar to that of a metal door.

For comparison: A single 1¾-inch wood door has an "R" factor of 2.04; when equipped with a wood storm door with a 50 percent glass area, the "R" factor is 3.70; and when equipped with a metal or glass storm door, the "R" factor is 3.03.

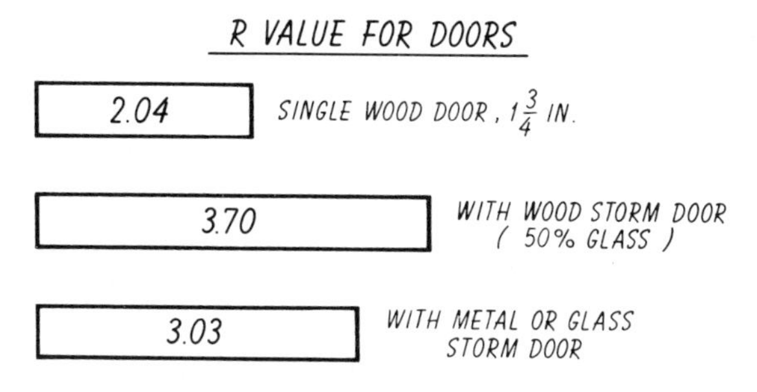

A spring-operated door closer on the storm door is a good investment. This is particularly true if it is strong enough to discourage loitering in the doorway. (As a rough estimate, each child and house-broken dog adds about 5 percent to the total winter fuel bill.)

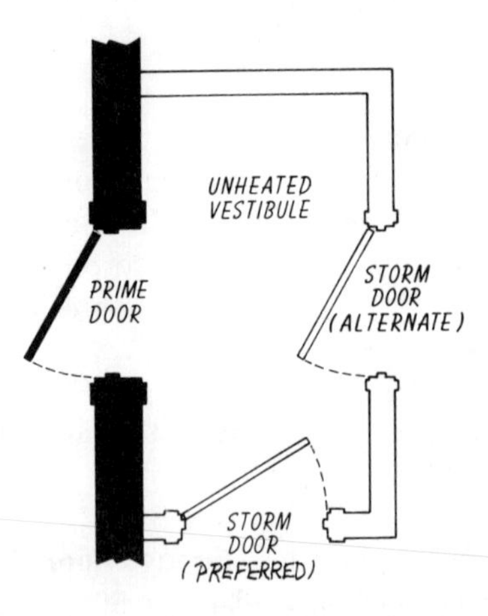

Since the lighter storm door is more subject to the force of wind, a chain or spring should be attached to the top of the door to prevent it from flying open and shattering the glass.

An acceptable alternative to a storm door and one that is particularly attractive to the elderly and handicapped, who do not wish to fight a storm door, is the metal prime door with a foam plastic core. A thermal break is installed around the edge of the door, and the doorframe is equipped with a magnetic weather strip similar to that used on refrigerators. This type of door, when closed, provides an insulating value fully equal to a conventional door with a storm door, but it does not prevent as much air exchange when open as the prime door–storm door combination.

Heat loss due to air infiltration will be greatly reduced if there is an unheated vestibule between the prime door and the storm door. This can often be created by enclosing a porch or a portion thereof. In northern climates temporary walls are often erected in the winter for this purpose.

Any exterior door that is not used during the heating season should be sealed tightly, perhaps even taped or caulked shut. This does not apply to emergency fire exits, of course. If the unused door is equipped with a storm door, some of the space between the two doors can be filled with blanket or batt insulation to provide additional protection. Doors leading to unheated portions of the house, such as attics and storage areas, should also be insulated and weather-stripped. Board-type insulation can be attached to the unfinished side of such doors.

4

Room Air Temperatures

Indoor design temperatures in the United States have increased from 70 degrees in the 1920s to 75 degrees in the early 1970s. In other words, the American's idea of a comfortable temperature has increased about 1 degree per decade. During this period the clothing worn indoors has become lighter, so that little difference exists between summer and winter attire.

METABOLISM AND HEAT LOSS

The role of clothing in relation to thermal comfort can be best understood by considering the heat processes of the human body. The body can be considered as a heat engine; the food we eat generates heat and provides energy for physical activity. Our body temperature is maintained at a remarkably constant level. The process that generates heat is called metabolism and the body must maintain this process and keep it in balance with heat loss if damage to the body is to be avoided.

The internal generation of heat depends upon the degree of physical activity performed. The metabolic rate can vary as widely as 350 Btuh for a person at rest to over 1,500 Btuh for a person doing heavy work. A person dusting, bedmaking, vacuum cleaning, and so on will probably show a metabolic rate on the order of 800 Btuh.

In order for the body temperature to remain constant, the heat loss from the body changes from minute to minute to adapt to changes in the environment. This heat loss from the body can take place in the four ways discussed earlier: convection, radiation, conduction, and evaporation.

FIVE DECADES OF INCREASE IN INDOOR TEMPERATURES

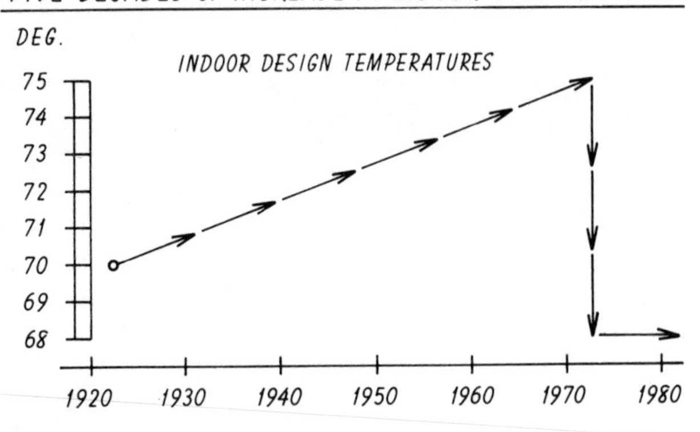

METABOLIC RATES

BTU PER HOUR

Activity	BTU per hour
SEATED AT REST	350 BTU PER ADULT
SEATED VERY LIGHT WORK	400
MODERATELY ACTIVE OFFICE WORK	450
STANDING, LIGHT WORK, WALKING SLOWLY	450 TO 500
SEDENTARY WORK (RESTAURANT)	550
LIGHT BENCH WORK (FACTORY)	750
MODERATE DANCING	850
WALKING 3 MILES PER HOUR	1000
HEAVY WORK	1450

Obviously, the heat loss from the body is affected by the type and weight of clothing. The human body has a remarkable temperature control system regulated by a nervous center located at the base of the brain, which can direct the many processes that maintain a constant temperature throughout the whole organism. This center is as busy as a symphony orchestra conductor, and, unless illness intervenes, it does its job for the entire life of the individual. All human beings maintain the same body temperature and are equally affected by their environments, except for small adjustments that their systems make to accommodate differences in diet, clothing, shelter, and climate.

As a person becomes chilled, he shivers, the heart beats faster, his muscles are caused to move, the blood flow to arms and legs is restricted, and every effort is made to conserve heat for the vital organs. The stomping and hand waving, as well as the compact body posture, of a chilled person are automatic responses to this sensitive control.

In a hot summer environment the response is the opposite. The muscles relax, the lungs operate faster to ventilate the body, the heart beats faster to keep the lungs moving and to move more blood to the arms and legs, the blood flows to the capillaries, which rise closer to the skin, and perspiration takes place to increase evaporative cooling. Again, the behavior is automatic, and a slowing of body movements and a less compact body posture are responses governed by the control center at the base of the brain.

If we expect to maintain the same thermal comfort at 68 degrees that we enjoyed at 75 degrees, some adjustment in our clothing will be necessary. The English, who have long been adjusted to 68 degrees or less, and suffer when they are first exposed to the American environment, wear clothing that has been adapted to their own climate.

For maximum fuel conservation, with no reduction in thermal comfort, a double standard for thermostat settings should be considered. A morning or early afternoon setting of 66 degrees can be comfortably endured if a person is physically active, as when doing household chores. After the chores are completed a setting of 68 degrees would be welcome, especially if the occupant is seated and at rest.

Two exceptions should be made to these lowered thermostat settings. Many older people, especially those who are not physically active or who have problems with blood circulation, may suffer discomfort. The body heat loss must be reduced, and this can be done by wearing an afghan around the legs, covering the neck, or using a heating pad to warm the hands. This also applies to young people who are ill.

Thermostat settings should not be "jiggled." When the setting is to be changed from 66 to 68 degrees, do not raise the setting to

80 degrees in the belief that the room will heat faster. The heat input to the furnace is fixed, and the rate at which the room heats up will remain the same whether the setting is raised 2 or 20 degrees. The objection to the 80 degree setting is that the occupant will forget to return the setting to its normal position until the house is overheated. Set the thermostat at the desired point, and then walk away from it.

Most people complain that a 68 degree thermostat setting does not provide the same warmth in very cold weather as it does in mild weather. This is to be expected, because room surfaces become colder and air temperatures in the lower part of the room are also colder. Complaints are more numerous from those people who live in poorly insulated houses or have poorly adjusted heating systems.

Another reason for the greater discomfort experienced in very cold weather is the thermostat itself. Most modern room thermostats have a small heating element *inside the case* in order to provide for more sensitive temperature control. (Normally, a variation of 1 degree above or below the desired setting is considered acceptable.) In cold weather, as the room thermostat is demanding heat for a longer period of time, the internal heating element tends to warm the thermostat at a faster rate than the rise in room air temperature. That is, the thermostat may become satisfied when the room temperature is still only 65 degrees, and it will show 68 degrees on its scale when the air in the room is actually colder.

Obtain a desk thermometer and place it on a table in the normal living area. Keep it away from the top of the television set or the area beneath a table lamp. In cold weather locate the correct thermostat setting by trial and error, until the desk thermometer in the room reads 68 degrees. This will save you an unnecessary service call to your heating man.

Since the heat loss from the arms and legs is the controlling factor in body heat loss, the use of long-sleeved garments and those that cover the legs will be helpful. Warmer, long-sleeved clothing for both men and women is likely to become the prevailing fashion over a longer seasonal period in the years to come.

Rugs and carpeting placed over tile, stone, and other conducting floor surfaces will reduce heat conduction through the soles of the shoe. Warm footwear will greatly reduce heat loss from this sensitive part of the body.

Most people, once accustomed to a 68 degree environment, report that they feel more alert than when they were in 75 degree environments. Over 75 degrees many people tend to become drowsy (as any college lecturer will verify).

The period of bodily adjustment between a summer temperature of 78 degrees and a winter temperature of 68 degrees takes some time. When the army trains soldiers to operate in a different climate, this adjustment period proves to be several weeks long. Many have experienced the sensation of feeling colder when that first freezing day in autumn is met than when the same sort of day occurs in spring. The body has become adjusted to the summer environment, and is reacting to autumn weather with greater sensitivity than it would to a cold day in early spring.

FUEL SAVINGS AT 68 DEGREES

For large parts of the United States with heavy concentrations of people (in the 4,500–6,500 degree-day zone), the average winter temperature is between 30 and 40 degrees. (This is the *average* of the entire heating season and not just the extreme cold months.) These temperatures correspond to an average indoor-outdoor temperature difference of between 35 and 45 degrees.

When the indoor air temperature is reduced 1 degree, the reduction in heat loss from the building will be between 2 and 3 percent. With a reduction in the indoor air temperature of 7 degrees (from 75 to 68 degrees), the corresponding reduction in the fuel requirement will be between 14 and 21 percent. In milder climates the fuel savings will tend toward the higher percentages.

In view of the larger percentage savings in milder weather, there is much to be said for maintaining a temperature of, say, 66 degrees or lower in fall and spring, and then raising the thermostat setting to 68 degrees in colder weather. This upward adjustment of temperature in colder weather makes sense because all the factors affecting comfort are worse in cold weather than in mild weather. For example, the air in the living area gets colder as the weather becomes colder (even with a fixed thermostat setting), the exposed walls and windows get colder, and even the drafts at the floor get colder and move faster. In spring, as the weather moderates, a thermostat setting of 66 degrees will feel just as comfortable as a 68 degree setting in cold weather.

A reduction of 7 degrees in the air temperature will result in an increase in the relative humidity of about 6 percent. This is desirable in the winter because indoor air tends to be too dry. This increase in humidity will occur without any additional release of moisture into the house and can be considered as a fringe benefit of reduced air temperatures.

Although 68 degrees has been established as a desirable thermostat setting, there is no magic in this number. It was set presumably to effect a reduction in energy consumption of about 15 percent. In English practice, as well as in many cool climates, even lower temperatures are accepted as normal.

After a period of adjustment to lower thermostat settings, personal adjustments can be made. For example, if both husband and wife are working at the same hours, the thermostat setting during their absence from the house could be set at 65 degrees, or perhaps even lower, and then turned up after they return home.

NIGHTTIME REDUCTION IN ROOM TEMPERATURES

Arguments rage as to whether any benefits occur with lowered thermostat settings at night. A number of thorough tests were conducted in several research residences at the University of Illinois under carefully monitored conditions and over a wide range of cold temperatures. The results showed that a large number of items can affect the potential fuel savings of lowering temperature settings at night.

The best way to explain this is to show three separate temperature graphs, each plotted against time.

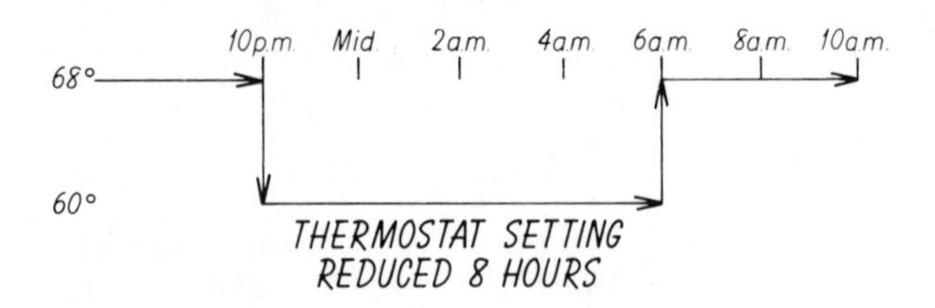

1) This graph shows that the thermostat was set at 68 degrees until 10:00 P.M. At that hour the setting was reduced to 60 degrees. In the morning at 6:00 A.M. the setting was increased to 68 degrees again.

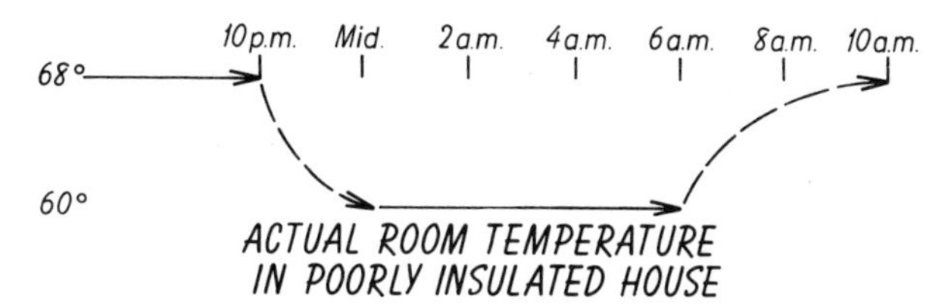

2) In the next graph the actual room air temperatures in a *poorly insulated* house are shown. The house cools rapidly in about 2 hours under these weather conditions and the temperature is then held at 60 degrees until 6:00 A.M. The recovery period is from 6:00 to 10:00 A.M. During this long recovery the furnace is operating at peak capacity and the flue gases leaving the chimney tend to get hotter and hotter.

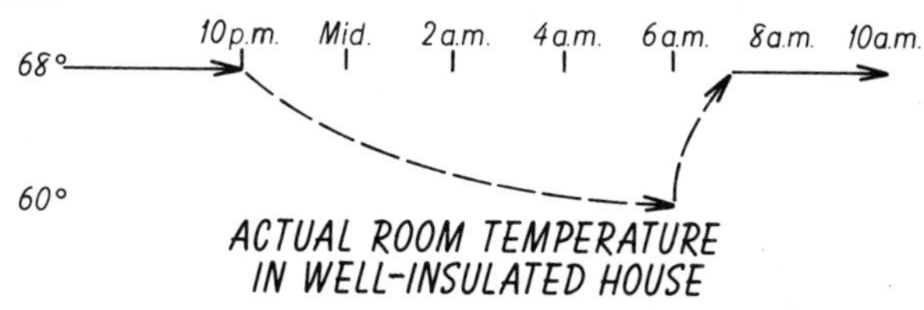

3) In the final graph the actual room air temperatures are shown for a *well-insulated* house under identical weather conditions. The room air temperatures never did fall to 60 degrees because the heat loss was small. In the recovery period the furnace operated for about 1 hour at peak capacity and then resumed normal operation.

The point of these three graphs is that many conditions can affect the fuel savings obtained from night thermostat reduction. In the many tests made under carefully controlled conditions, the following conclusions were reached:

For a 10 degree night thermostat reduction, the observed fuel savings ranged from zero to as high as 15 percent, with an average savings of about 7 percent.

The long recovery period in the morning can increase the wear and tear on your furnace equipment. For this reason large night temperature reductions are not recommended in extremely cold weather. At such times the fuel saving will be small because the heat loss through the chimney will be high. Also, furniture, walls, floors, and other objects will be cold to the touch and will not reach 68 degrees for some time after the setting has been moved upward again.

Some people prefer to have automatic day and night thermostat settings. A clock thermostat with adjustable settings for day and night temperatures is available. This device can reduce the setting at 10 P.M. and restore it about one hour before normal waking time.

ELECTRIC BLANKETS AND SPOT HEATERS

Sleeping at night in temperatures of 60 degrees or less (especially for those who insist on opening a window) can be made tolerable by the use of electric blankets. Some critics and rugged types have scoffed at the idea, and explained that such devices use electrical energy, and that extra blankets would be equally effective.

Without going into the relative merits of featherbeds, sleeping bags, wool blankets, quilts, cotton blankets, and so on, there is a legitimate argument in favor of the electric blanket. If one sleeps better with lighter covering, this alone is a good reason. Basically, the decision is between the use of about 175 to 200 watts on an intermittent basis and the much larger energy savings that result from lowering the entire house temperature and applying heat directly to the bed.

The shock of stepping out of a warm bed into a cold environment can be reduced by means of spot heaters. For example, infrared lamps are often installed in bathrooms so that heat radiation can be focused on the person below. Some of these lamps are operated from a time switch so that the lamp is turned off after a set number of minutes. This is the modern equivalent of the experiences of many old-timers who recall their childhood days of rushing to the glowing parlor stove and getting dressed with inner garments first exposed to the radiant heat. Unfortunately, the mother who started the fire had no spot heater to toast her.

SUMMER INDOOR TEMPERATURE

In current practice a year-round indoor temperature of 75 degrees has been accepted as a design value. For the purpose of reducing energy requirements for summer cooling, a value of about 78 degrees has been proposed as a suitable summer indoor temperature. It is interesting to note that the first extensive tests on residential summer cooling made at the University of Illinois in the 1930s settled on 78 degrees as a reasonable value, just below the threshold of being too warm.

SUMMER
INDOOR
TEMPERATURES ——— 78 DEG.

SUMMER
INDOOR
RELATIVE HUMIDITY——NOT OVER 60%

An indoor temperature of 78 degrees will be acceptable for summer provided that the relative humidity can be maintained at less than 60 percent. In dry climates this may be readily accomplished, but in humid areas it may be difficult to obtain humidities below 60 percent without also reducing air temperatures to below 78 degrees. Local experience will dictate a practical compromise that is just on the borderline of discomfort.

5

Can a Humidifier Save Fuel?

HUMIDITY AND THERMAL COMFORT

In early research studies, extending back some fifty years, investigators noted that young persons were just as comfortable in an environment that was a few degrees cooler than normal, provided that the relative humidity was increased a few percent. At the time the conclusion was reached that if the relative humidity could be increased by about 10 percent, the indoor temperature could be reduced by about 1 degree with the same feeling of thermal comfort. These pioneering tests were conducted by having test subjects move from one experimental chamber to another. In other words, the occupancy period was short, as it might be if a person moved from one room to the next and never stayed in any one location for more than a few minutes.

In the 1960s additional experiments were conducted for long-term occupancy of several hours in a particular test environment. The results this time did not confirm the early tests—in fact, relative humidity was not found to be a factor except in cases of extremely dry or humid air. The only factor that was critical, as far as environmental comfort was concerned, was the dry-bulb temperature. The occupant of a home usually stays indoors for long periods. The conclusions of the tests in the 1960s, therefore, are based on a much more realistic set of conditions.

However, old traditions die slowly. Today many news articles and a great deal of advertising literature repeat the now obsolete notion that increasing the relative humidity will result in fuel conservation. In fact, *one pays for increased relative humidity.*

The evaporation of water that increases the relative humidity requires

energy—over 1,050 Btu for every pound of water evaporated. Since 1 gallon of water contains 8.3 pounds, this means that the evaporation of 1 gallon of water requires almost 9,000 Btu. In an ordinary house, which requires from 3 to 8 gallons of water evaporation to maintain from 30 to 40 percent relative humidity in cold weather, the energy requirement can be substantial. If the water is evaporated at room temperature, the heat required comes from the room air. If electricity is used to heat the water, the energy comes both from the electric current and from the air. In general, therefore, any increase in relative humidity will not save energy.

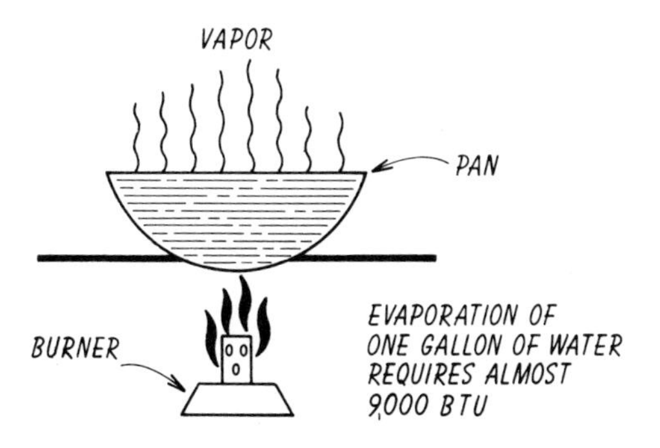

EXTREMELY LOW HUMIDITY CONDITIONS

There are some positive aspects of humidification, and these need to be stated so that the homeowner has a correct overall picture of the role of relative humidity. The main objections to extremely low relative humidities can be summarized as follows:

a) Organic materials shrink as moisture is lost to the dry air. These materials include wood, wool, hair, cotton, most textiles, leather, and products made of these materials.

b) Many people complain of dry nasal passages and dry throats.

c) Quick drying of skin surfaces causes chapping and scaling of skin.

d) Charges of static electricity build up easily and can prove extremely annoying to the occupants of houses with certain types of carpets and rugs.

e) Dust particles tend to stay in suspension in the air instead of settling.

One further reason for maintaining higher indoor relative humidities is the conclusion of a report of experiments conducted at a medical school where bacterial growth was minimized at the extremely high relative humidity of 50 percent. But until additional evidence is obtained, and the effect of humidity on harmful organisms is clearly demonstrated, this point must be considered unproven.

In any case it seems that there is some justification for maintaining a *reasonable* relative humidity in the winter.

What is a reasonable humidity? The relative humidity in a house depends upon the weather. In general, the humidity becomes lower as the outdoor temperature drops. In fact, when the outdoor temperature drops to 0 degrees, the indoor humidity can decrease to as little as 10 percent. The reason that it is usually somewhat higher, even in cold weather, is that moisture is being released in the house from one or more of the following sources:

a) bathing and showers
b) cooking, baking, and boiling
c) laundry washing, drying, and ironing
d) indoor plants
e) humidifiers

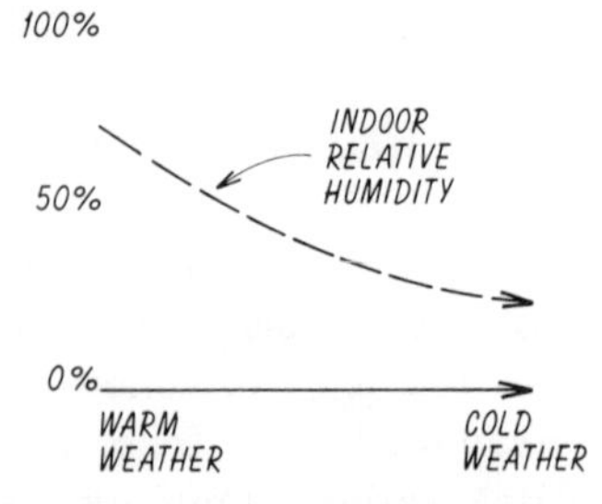

An extremely low relative humidity is caused when normal infiltration and ventilation bring in outdoor air (with its low specific moisture content) so that an equal amount of room air (with its higher moisture content) is forced out of the house at some other point.

One of the first requirements for retaining moisture generated in the house is to reduce air infiltration and ventilation. This is one of the fringe benefits of using weather stripping around doors and windows and of using storm doors and storm sash.

Kitchen and bathroom ventilators should be used sparingly in cold weather, mainly to remove odors. Be sure the automatic duct dampers

are operating and closing tightly. If air leaks occur, they might be sealed shut for the winter.

Try to eliminate activities that produce odors. Smoke less and avoid cooking odorous foods, thereby reducing the necessity to ventilate the house.

Whatever moisture is released in the bathroom through bathing and showering should be allowed to dissipate into the rest of the house by opening the bathroom door soon after use, rather than operating the bathroom vent fan.

Use static eliminator sprays on rugs where static shock is a problem or select carpeting with fine stainless steel wires woven into it to ground any static charge.

Keep the door to the outside closed as much as possible in cold weather to reduce air infiltration. Children should be cautioned to "stay in, stay out, but not half in and half out." The installation of automatic door closers will be helpful.

The use of a temporary or permanent vestibule, with a separate inner door, also reduces infiltration considerably.

Normally, indoor relative humidities are too low rather than too high, and even with all the moisture release that occurs indoors the homeowner still complains of dry throat, chapped skin, and dry nasal passages. Sometimes the drying effects on musical instruments made of wood (pianos and violins in particular) are considered harmful. The last resort of the homeowner is to purchase a humidifier.

The purchase of a desk-type humidity indicator is recommended if one is not attached to the humidifier. Although these instruments are usually not too accurate, they do provide two reference points that can be useful. Regardless of what the humidity indicator reads, mark the point on the scale where you notice discomfort due to drying of nasal passages or the throat. This will be the point where the humidifier should begin operating. Mark also the upper point where the symptoms seem to disappear (or where static electricity is not being formed). Shut off the humidifier in mild weather as soon as the indicator shows a satisfactory level.

EXTREMELY HIGH RELATIVE HUMIDITY

Unfortunately, for the number of houses that seem to be too dry there seem to be an equal number that show symptoms of too much indoor humidity. The main indication of excessive moisture in the house is fogging or frost on a window surface. If the humidity is so high that condensation occurs on the inside surface of a window with insulating glass or a storm panel, condensation is probably occurring inside walls, and structural damage may result. The humidity should be reduced as soon as possible. Single-glazed windows fog at a much lower relative humidity than double glazing. For this reason very few commercial, public, or institutional buildings can be expected to maintain even 25 percent relative humidity in very cold weather. If window condensation is a serious problem, the first requirement is to reduce moisture input.

In this case the procedures previously mentioned are reversed. Obviously, the humidifier must be shut off.

Kitchen and bathroom vent fans should be turned on after moisture release. Windows or outside doors can be opened for a while.

Laundry dryers should be vented to the outdoors.

Be sure there is a plastic vapor barrier or concrete slab over the earth in any crawl space. A bare earth surface acts like a giant humidifier, even when the surface appears dry.

Install storm sash, double glazing, or a temporary substitute on all single-glazed windows.

Check on all sources of moisture input inside the house.

Normally, during a winter heating season it is possible to maintain a fairly high indoor humidity (50 percent) when the outdoor air temperature falls between 50 and 60 degrees. As the weather becomes colder, the humidity should be reduced—it will be reduced automatically unless a humidifier is used. For outdoor temperatures between 20 and 40 degrees, a humidity of 40 percent is reasonable for houses with storm sash and 20 percent for houses with single glazing. In zero weather another 10 percent reduction in humidity might be necessary to avoid window condensation.

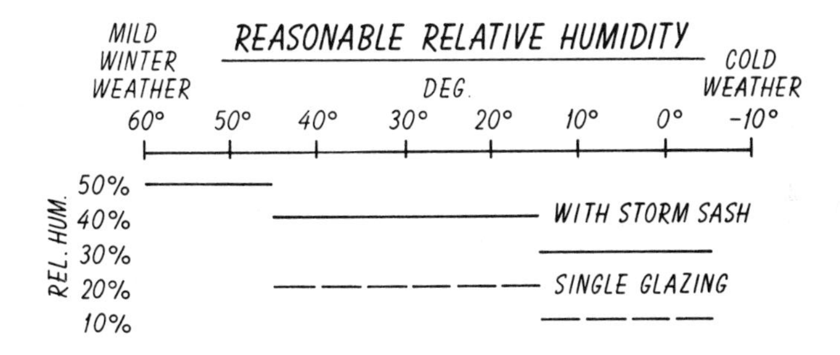

For energy conservation, and in order to avoid frequent adjustments of the humidifier control, the relative humidity should be maintained at the lowest setting necessary for your physical comfort.

SUMMER HUMIDITY CONDITIONS

Maintaining proper relative humidity in the summer is an entirely different problem. In the arid areas of the Southwest, the summer indoor humidity presents no difficulties.

In a large part of the country, however, the indoor humidity in the summer is too high. Under extreme conditions the high humidity promotes mold growth on leather, damp wood, damp soil, and so on, and causes rusting of steel and iron. In these areas the use of a dehumidifier for basements or enclosed spaces is almost a necessity.

One of the major benefits of mechanical summer cooling, that is, air conditioning, is the reduction of relative humidity that accompanies the operation of the cooling unit. The finer points of cooling unit operation are described in a later chapter.

The use of a dehumidifier may permit the indoor humidity to be maintained at about 70 percent or below even in damp, muggy weather.

The basement space or enclosure should be protected as much as possible from frequent openings of doors and windows that allow outdoor air and moisture to enter the area.

In some basements the water vapor is supplied by basement walls that are somewhat damp. As far as possible the earth around the house should be sloped to divert water away from the foundation. This may require the filling in of low spots with tightly packed earth.

Gutters and downspouts should be used to collect and dispose of roof runoff. The downspout should discharge at least 5 feet away from the foundation and at a point where there is an obvious slope away from the house or into a drainage tile. Downspouts should never be connected to the tile around the bottom of the foundation.

There is a commercial treatment for damp walls that consists of an application of sealer on the outside of the basement wall.

If walls are damp, but not leaking water, a brush-applied sealer can reduce the moisture problem significantly. The product is available at paint stores.

The basement floor can be painted with concrete floor enamel. This will seal the pores of the concrete. The paint should be applied in the winter when the concrete is dry.

Concrete stepping-stones (placed close together and preferably cemented to each other) can be placed along the foundation to serve as a watershed.

Basement access areas and window wells can collect water during heavy rainstorms. A glass or plastic cover over these spaces will divert rainwater. (Be sure to keep animals from stepping on the glass.)

In severe cases of water seepage through basement walls, it is possible that the footing drain is clogged and not carrying water away from the foundation. This may require removal of tree roots or actual replacement of drain tiles.

6

Fuels and Electricity

The common fuels, also called "fossil fuels," are coal, fuel oil, natural gas, and propane (or liquefied petroleum gas). Wood might be added to this list, though it is not a fossil product, and is no longer a common fuel. Each of these fuels contains hydrogen and carbon in their chemical makeup, and so they are sometimes known as hydrocarbons. Both hydrogen and carbon, when burned in the presence of air (which contains about 21 percent oxygen), release large amounts of heat energy.

For example, hydrogen burns with air to form water vapor and releases over 61,000 Btu for each pound of hydrogen burned. Carbon burns with air to form carbon dioxide and releases over 14,000 Btu for each pound of carbon burned. When too small a supply of air is combined with carbon in combustion, the dangerous gas carbon monoxide is formed. In any kind of combustion the complete oxidization of carbon, its conversion to carbon dioxide, is the desired goal.

The burning process requires large amounts of air. Normally, the infiltration of air into a house is sufficient to meet these requirements. In special cases, however, such as in a furnace room or furnace closet located in the living area, special provisions must be made to bring outdoor air into the furnace room. An air duct extending to a well-ventilated attic is often used. The products of combustion from the burning of fossil fuels can contain a wide variety of gases, some of them objectionable and some of them dangerous to health.

These gases include carbon dioxide, carbon monoxide, oxygen, nitrogen, nitrogen oxide, water vapor, and sulfur dioxide. These gases are dumped into the atmosphere through the flue and are dispersed by the wind.

Because the burning of coal results in the formation of sulfur dioxide as well as carbon monoxide, coal is not a desirable fuel for home-heating equipment. The preferred fuels for home heating are fuel oil, gas, propane, and electricity (which should be classified as a fuel for our purposes).

The good performance of any heating device depends upon the following conditions

1) The lowest fuel input that will still provide for some margin of capacity (at least 10 percent) during the coldest day.

2) The proper amount of air for good combustion. Too much air is wasteful of heat and too little could result in the formation of soot and carbon monoxide.

3) A proper mixture of fuel and air.

4) A sufficiently high temperature during the burning process so that the flame is not chilled before the reaction is complete.

An experienced serviceman can almost always find the right combination of adjustments by merely looking at the fire. A correctly adjusted gas flame is blue white and not noisy; an oil flame is yellow orange with a hint of smokiness at the tip. In all cases instruments would show that the carbon dioxide content of the flue gas was relatively high.

One basic instrument used by a serviceman is a flue gas thermometer that will read up to about 1,000 degrees. A high flue-gas temperature indicates that the loss through the chimney is high at the point of measurement.

A very sharp drop in flue-gas temperature occurs between the furnace and the inlet to the chimney. This loss in heat is regained by the basement or furnace room, as is the heat absorbed by the chimney. This heat is just as good as the heat delivered to the rooms upstairs, and should be conserved just as carefully. Any heat released into the basement warms the floors of the rooms above and provides for better thermal comfort.

If, on a very cold night, when the heating system is supposed to be operating almost continuously, the blower or circulating pump and burner operate only infrequently and with long off periods, you might jot down the times when the burner starts and stops. Ask your serviceman if the burner input might be too large for the house. Keep in mind that if the house temperature has been reduced by 7 degrees, this alone reduces the load by about 14 to 21 percent. If the burner input is reduced, fuel will be saved. Do not attempt to make this adjustment yourself. Some burners will not work properly with fuel input rates much below approved settings. The input rate affects the efficiency of the fuel's ignition and the completeness of combustion.

Tests have shown that the ordinary domestic heating system is an extremely efficient system. It is important at this point to introduce the concept of "overall heating efficiency." Not all heat losses from a furnace are detrimental. The heat lost from the house and the heat lost through the chimney are not regained and represent a waste of fuel. The so-called heat losses from the equipment—through the casing, bonnet, ducts or pipes, flue pipe, and even the chimney walls—are all regained by the house. This vagrant heat really ends up as useful heat and represents an efficient use of fuel.

Average values for overall heating efficiency are shown in the table below. The prices quoted are approximations of the costs of various fuels at the time of writing, and while this table will give you an idea of the relative efficiency of these fuels in terms of their current availability and price structure, prices are of course continually fluctuating, and the relationship between efficiency and cost is constantly changing. The reader will himself have to recalculate these values at the time he considers an alternate source of energy for his home.

$$\frac{\text{(Btu value of fuel per unit of sale)} \times \text{(overall efficiency)}}{\text{(cost in cents per unit of sale)}}$$

Let's show four examples:

1) No. 2 fuel oil sells at 25 cents per gallon:

$$\frac{(140{,}000)\ (0.80)}{25} = 4{,}480 \text{ Btu for 1 cent}$$

2) Natural gas sells for 12 cents per therm

$$\frac{(100{,}000)\ (0.80)}{12} = 6{,}667 \text{ Btu for 1 cent}$$

3) Propane sells for 40 cents per gallon:

$$\frac{(91{,}000)\ (0.80)}{40} = 1{,}820 \text{ Btu for 1 cent}$$

4) Electricity sells for 2 cents per kilowatt-hour:

$$\frac{(3{,}413)\ (1.0)}{2} = 1{,}706 \text{ Btu for 1 cent}$$

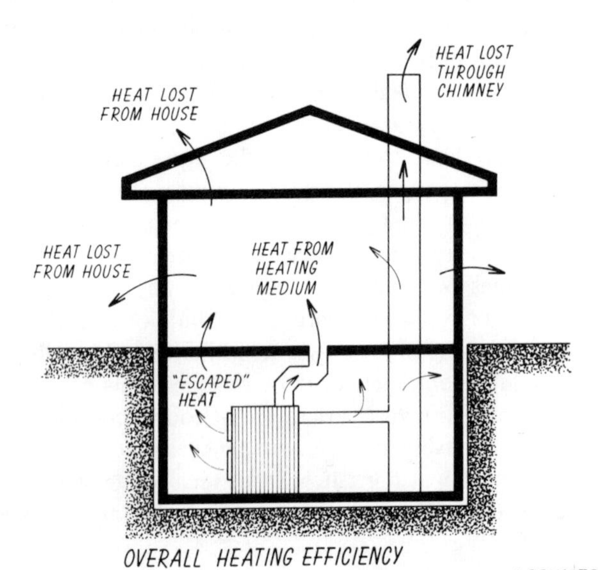

OVERALL HEATING EFFICIENCY

TYPE OF FUEL	METHOD OF BURNING	ASSUMED OVERALL EFFICIENCY
BITUMINOUS COAL	STOKER-FIRED	65 %
OIL	UNITS DESIGNED FOR OIL BURNING WITH EXCEPTION OF VAPORIZING TYPE WITHOUT FAN	80 %
OIL	VAPORIZING-TYPE UNITS WITHOUT FAN, AND UNITS CONVERTED TO OIL BURNING.	70 %
GAS (ALL TYPES)	ANY PROPERLY DESIGNED BURNER	80 %
ELECTRIC (IN ROOM)*	CEILING CABLE, BASEBOARDS	100 %
ELECTRIC (CENTRAL)*	DUCTED FURNACE, BOILER	90 %

* WHILE ELECTRICAL CURRENT IS NOT A FUEL, IT IS SOMETIMES USED TO SUPPLY HEAT AND FOR THIS REASON IS INCLUDED IN THE TABLE FOR COMPARISON PURPOSES.

FUEL OIL AND OIL BURNERS

Fuel oil for home heating is sold in two grades: No. 1 (which looks like kerosene) has a heating value of 136,000 Btu per gallon. This fuel is primarily burned in vaporizing (pot-type) burners; No. 2 fuel oil is heavier and contains about 140,000 Btu per gallon, and is used in pressure-type or gun burners. Both fuels are relatively safe to handle and can be stored in indoor tanks. Fire protection codes limit the amount that can be stored indoors (usually two 275-gallon tanks), and local requirements should be obtained from the building inspector.

The most common burner is the pressure-type or gun burner, which contains a nozzle through which the oil is injected under high pressure. Most service problems arise in connection with this nozzle, which has an opening so small that impurities in the oil can lodge in it and seriously affect the way in which the oil spray is formed. A spark ignites the oil spray in the presence of a swirl of air and, once ignited, the flame continues without further need for the spark.

The cleaning of the oil filter and the spray nozzle is not a difficult task and oil burner users should become familiar with the process. Be sure not to use any hard material to clean the nozzle—a wooden match stick is as hard as one dare use on this delicate device.

If sooty combustion is occurring—and this is common when the nozzle has been plugged—the soot should be brushed or blown off the control equipment placed in the path of the flue gas.

NATURAL GAS

Natural gas, or methane, is odorless and half as heavy as air, so that if leaks occur it tends to rise to the ceiling. A distinctive odor, mercaptan, is added to the gas to serve as a warning in case of leakage. Gas pipe installations should be made by competent men who realize the possible dangers from poor installation.

Most natural gas has a heating value of about 1,000 Btu for each cubic foot burned. The fuel industry has devised a measure called a "therm," which is equivalent to 100,000 Btu or approximately 100 cubic feet of gas. The ordinary gas meter records the flow of gas in cubic feet. However, since the heating value of gas could vary from day to day, or from one area to the next, the fairest way to treat a customer is to convert the cubic feet meter readings to therms. The gas billings are in terms of therms sold to a customer.

There is little that can go wrong with a gas burner once the gas input to the orifice inside the burner is adjusted and the air openings are set and locked into position. In general, a hands-off policy is best for the homeowner. Leave any problems that might arise to the serviceman.

PROPANE GAS (LPG)

Liquefied petroleum gas can be pure propane or a mixture of propane and butane, and it is delivered in a heavy pressurized tank truck. The liquid is forced under pressure into the home storage tank, which should always be located *outside the house.* The liquid in the tank is changed to a gas as a result of the warming action of the outdoor air, and this gas is then piped indoors and handled like natural gas.

There is a great difference between propane and natural gas. Both propane and butane are about twice as heavy as air. Consequently, when released into the air LPG tends to sink toward the ground. In the event of a leak or an ignition failure with a propane burner, gas will collect at floor level and spread outward like any dense liquid.

Propane equipment should be used only with full knowledge of the potential hazards. The orifice size and the placement of the pilot light for a propane burner are not the same as those of a natural gas burner, so that before any conversion is made, competent servicemen should be called. Propane contains almost twice the heating value of natural gas per cubic foot of gas, so that if no adjustment of the fuel flow is made when converting, an oversized flame will result.

Propane liquid is sold by the gallon or pound and contains about 91,000 Btu per gallon. Pure butane contains about 103,000 Btu per gallon. The product sold in your area may be a mixture of these gases.

LIQUEFIED NATURAL GAS (LNG)

Natural gas obtained from a foreign country must be liquefied first at the well site. This can be done by cooling it to about −160 degrees Fahrenheit. It is then poured into special "thermos bottle" container ships and delivered to some coastal city to be stored in liquid form. When natural gas is needed, the liquid is pumped into pipes that are exposed to heat and it boils off in the form of gas. After this it can be handled like any other natural gas. Though not now common, the fuel shortage will increase its use.

ELECTRICITY

Electrical energy is often derived from the burning of fuel, and replaces other fuels in the home-heating system, so it is included here.

When electrical energy is converted to heat energy, 3,413 Btu are obtained for each kilowatt-hour. In most instances the conversion of energy occurs at an efficiency of 100 percent at the *point of application*. There is a distinction to be made between efficiency at the point of application and the overall efficiency of the total electrical system.

For example, at the electrical generating station pulverized coal is burned in large boilers at high temperatures, so high that the ash in the coal melts and forms a slag, The heat concentration is great, which enables a ten-story high boiler to burn a carload of coal each hour. The cost of generating this heat is less when high heat concentrations are possible. A substantial part of this thermal energy is used to heat water and change it into high-pressure steam. The steam is delivered to huge turbines, which turn the generators and produce electricity.

The flue gases in these huge boilers are extremely hot, and even after passing over the boiler tubes they are so hot that it would be wasteful to throw the gases away. Devices are installed in the flue gas stream to extract heat for the water that is fed to the boiler and to heat the air that is used for combustion. Finally, the still-hot flue gases are directed up the stack to be dumped.

Vented flue gases are not the only source of thermal waste in a power plant. The steam that passed through the turbine is still hot. Since steam cannot be pumped back into the boiler, it has to be condensed by cooling. Eventually, therefore, a large quantity of heat is thrown away through the condenser and the cooling tower.

A modern power plant is able to produce about 1 kilowatt-hour from about ¾ pound of coal. This amounts to a boiler-plant efficiency of about 35 to 38 percent. Since additional losses occur in the transformers and the transmission lines between the power plant and the home, the system efficiency is approximately 30 percent.

Is an electrical heating system that is 100 percent efficient in the home really 100 percent efficient or only about 30 percent efficient? In terms of the effective utilization of energy in the original coal, the 30 percent value is the only value that makes sense.

One other aspect of electrical heating needs to be brought out. An electric light bulb gives off light energy and also some heat energy. Eventually, the light energy is absorbed by the walls, floor, and furnishings in the room and turned into heat energy. In other words, the electric light bulb is itself a crude type of heating device. So also is the television set and all motors in the house.

To some extent all American homes are electrically heated. Some might argue that if the use of lights and appliances is reduced, then less heat will be generated by electricity, and more gas or oil will be required. It is hard to deny this fact.

There are many points that have to be considered here. If electricity is generated by burning fuel oil at the power plant, for example, then the electrical heating of residences is definitely not as economical a use of oil as burning oil in a home furnace. On the other hand, electricity generated by the burning of coal or by nuclear materials is less wasteful than burning fuel oil in a home-heating system.

This leads us back to the three distinct ways in which electrical energy can be used for home heating:

1) by resistance heating (passing a current through a resistance element)

2) by radiant spot heaters, as described in the chapter "Room Air Temperatures"

3) by a heat pump, as will be described later.

Basically, electricity can be used in all the ways that fossil fuels can be used, and many more.

1) Electrical resistance elements can be immersed in a steam or hot-water boiler.

2) Resistance coils can be placed inside a warm air furnace or inserted in an air supply duct.

3) Heating elements can be placed in a ceiling or in panels that are hung on a sidewall or ceiling.

4) Heating elements can be placed inside radiators that are filled with water, but have no piping connections.

5) Coils can be placed in a baseboard radiator.

6) Heating elements can be placed in a floor if the floor material is a good conductor of heat.

THE FUTURE OF ENERGY SOURCES FOR HEATING

During the past fifty years the American economy has undergone an energy revolution of major proportions without much public fanfare until recently. In the 1920s the major source of home-heating fuel was coal, with wood used in some areas and manufactured gas in a few cities. The hand-firing of coal was partly replaced by stoker firing, and later by oil burners.

After World War II, when the Big Inch was converted to a gas pipeline, the era of natural gas began. The conversion to natural gas-burning equipment was made at an astounding rate.

In the 1960s many electrical heating installations made their appearance. Then came the energy crunch of the 1970s. The question that now comes to everyone's mind is: "What will be the main source of energy for home heating in the future?"

The solution hinges upon a number of alternative decisions that might be made in the marketplace or in government. For example, if liquid fuel (oil) should be reserved for transportation needs and not for burning, then its use as a home-heating fuel might well decline. Similarly, if propane fuel is reserved for areas where natural gas cannot be used, its use as a heating fuel could easily disappear completely.

If the price of natural gas should increase sharply because of depletion of domestic sources, and a large portion was imported as LNG or derived from coal, then its cost would become closer to that of electricity in dollars per equivalent therm. If at the same time the production of electricity from nuclear fuels increased and the price increase was not as sharp as that of natural gas, then it is possible that an almost total conversion to electricity as a means of home heating could take place. Perhaps the heat pump would tip the scale in favor of electricity. *Perhaps all future housing should be built as if the source of heating were to be electricity.*

If that day should ever come, the problem of choosing a source of energy would not be as difficult at it is now, because there would be fewer options open to us.

7

All about Fireplaces

The English know more about fireplaces than any other people. The fireplace has been a basic means of heating in cool damp climates that seldom experience subfreezing temperatures. Early settlers brought the fireplace to this country and many models can be seen in Colonial Williamsburg in Virginia. One of the most impressive sights in the Governor's Palace at Williamsburg are the fireplaces in each of the major rooms. In the days before stoves and central heating the fireplace was the sole means of heating the home.

In modern homes the fireplace is an aesthetic adornment, and its function as a heating device has become of secondary importance. Occasionally, during prolonged ice storms that bring down power lines, the fireplace can be used to provide heat until the emergency is over. In general, the efficiency of a fireplace is low, so that its use has been limited to emergencies or to provide a cheery atmosphere.

The fire in the fireplace gives off most of its heat by radiation from the flames, the hot coals, and the surface at the back of the fireplace. However, a large amount of heated room air is drawn up the chimney and vented to the outdoors, and this is a waste of thermal energy.

When the fireplace chimney is located on an inside wall, part of the heat absorbed by the chimney is recovered by the house. However, if the fireplace chimney is located on an outside wall, so that three sides of the chimney are exposed to the outdoors, very little heat from the chimney is regained by the interior of the house.

As a safety measure, when prolonged use is made of the fireplace, combustible material should not be in close proximity to the rear

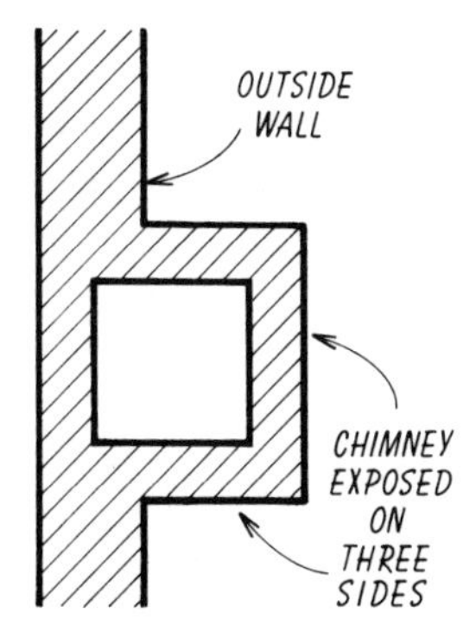

surface of the fireplace, since the transmitted heat can char wood or blister the painted surface of any object touching the back wall.

The efficiency of an ordinary fireplase is only about 10 to 15 percent. Therefore, unless firewood is plentiful and cheap, the constant use of a fireplace may be extremely costly.

The efficiency of a prefabricated fireplace that provides a means for circulating room air over the back and side heating surfaces is considerably higher than that of the ordinary fireplace. Information about these prefabricated units can be obtained from the manufacturers. In some models a fan is built into the unit to force heated air toward the lower part of the room where it does the most good.

Gas-fired logs in which the flue gases are vented up the chimney are inefficient burners of gas compared to boilers or furnaces.

An electric log, which is nothing more than an electric light designed to resemble a fireplace, will be no more wasteful of electric energy than an ordinary light bulb, as long as the chimney is closed or blocked. If the flue is left open as an exhaust vent, it should be closed as soon as the need for venting is accomplished.

The main advantage of a fireplace is that a person can sit or stand in front of the fire and endure reduced room air temperatures without discomfort.

A person seated in a high-back wing chair that completely encloses the body on three sides will lose little heat through the enclosed sides. When the seat faces the fire, the exposed portion of the body is warmed by radiant energy. This is a very effective use of radiant heating. The temperature of the glowing embers will be greater than 1,000 degrees.

When the fireplace is located in the same room as the room thermostat for the central heating system, the heat supplied by the fire will tend to keep the room warm enough so that the thermostat does not call for heat. Under these conditions the remainder of the house will be cooler than normal. A further saving in fuel will result because of the underheating of the rooms not exposed to the heat from the fireplace.

This suggests the need to gather the family in one room that can be maintained at comfortable temperatures. Such a warm haven could be the room containing the fireplace, as well as the television set, which is also a source of heat.

After the fire has died down, the warm room air will continue to escape up the chimney. Since the fireplace will continue to release combustion gases as long as the embers are warm, and the damper cannot be closed until the fire is dead, some means must be found to close the front of the fireplace.

Attractive covers for the fireplace, made of tempered glass, are available. The glass doors can be closed even with the fire still active. A low fire can be left to die out without attention, and the fireplace damper can be closed in the morning.

The use of a cover made from a fire-resistant material, such as pressed asbestos board, is also possible.

After the fire has died down, and the embers are completely cold, the flue damper should be closed.

Where environmental considerations do not prohibit the burning of coal, it may be used in place of firewood. Some coals contain considerable sulfur so that the combustion process results in the formation of sulfur dioxide, which is an obnoxious and toxic gas. Information regarding the sulfur content of coal can be obtained from local retail suppliers. The main advantage of coal as a fireplace fuel is that the fire does not have to be replenished as often as with wood and it is usually free from flying sparks.

From the standpoint of the efficient use of fuel energy, coal can be more effectively burned in a furnace or boiler.

For emergency operation coal-burning stoves have been placed in front of the fireplace, and the flue pipe from the stove run into the fireplace opening. A sheet-metal cover placed over the fireplace front ensures a stronger draft for the stove. Obviously, this requires a happy combination of a squat-looking stove and a fairly high fireplace opening.

Care should be taken to make certain that a coal fire is extinguished before the flue damper is closed, since the possibility exists that carbon monoxide may be formed by the dying embers, and this gas should never be allowed to enter the house.

In homes where paper is not being recycled and is merely dumped into the refuse can, the use of this paper for the fireplace should be considered. Papers are folded in half and stacked until they are about 1 inch thick. They are then tightly rolled to make a simulated log. Metal plant ties (or plain pliable wire) are wrapped around the bundle so that the "log" does not fly open when the outer layer is exposed to the flame. Using old newspapers in this way is not to be preferred to recycling waste paper, but it is better than throwing the paper into the refuse can.

It is possible to burn cartons, wood crates, mill ends, and other materials in a fireplace, but these should be handled with care because of the possibility of overfiring. The main drawback to the use of various kinds of waste material is that constant firing is required and the fire cannot be left unattended.

8

Residential Heating Systems

In those sections of the United States where heating systems are required to operate for more than six months of the year, central heating systems have become very popular. In the warmer sections of the country with a seasonal degree-day total of less than 2,000 to 3,000, room heaters are common. The discussion here will focus primarily on central heating systems, a distinct American contribution to better living.

Contrary to the statements made by some people, the domestic central heating system is very efficient when properly selected and installed by competent dealers. The statement is frequently made that in a certain town or county a central heating station delivers hot water or steam to a large number of nearby homes and that this is a more efficient method than generating heat in individual homes. Tales of this sort do not take into consideration the complete accounting of heat generation and distribution. From the standpoint of most efficient energy utilization, it is difficult to surpass a domestic system, in which heat is generated in a furnace or boiler, and any so-called waste heat (from casing, ducts, pipes, and chimney) is largely regained and utilized by the house.

In the first place an approved gas-burning furnace or boiler design must pass rigid laboratory tests by the American Gas Association and have a unit efficiency of at least 75 percent, as well as meet dozens of strict safety requirements. When the recovery of "waste heat" (called vagrant heat) is taken into account, the overall utilization of fuel energy in a home ranges between 80 and 90 percent.

This is far beyond the capabilities of any large central heating plant, even with professional firemen using sophisticated instruments.

The heat discharged from the main stacks is not recovered, large transmission losses occur through the pipes to the various buildings, and part of the water or steam is lost in the circuit.

Having disposed of the myth, this does not mean that *all* central systems are properly selected, properly installed, or properly maintained. This discussion, therefore, will seem to emphasize some of the deficiencies in central heating systems more than it does their advantages.

One other point needs to be made. If a central heating unit does not maintain uniform temperatures in each room of a house, it is known as an *unbalanced* system. It is common to find that such a system is operated at peak loads too often in order to take care of a hard-to-heat room. Under these conditions the unit will, of course, be wasting fuel.

Any changes that will provide more uniform temperatures, and particularly improved temperatures in the lower part of the rooms, make the house more comfortable.

In general, if the system can be operated at a low level of heat input and more continuously, it will be more efficient in its use of energy.

STEAM HEATING SYSTEMS

Steam heating systems have largely disappeared from homes because other systems are less costly and provide better temperature control. The steam generated in a boiler flows to the radiators by steam pressure created by the burner. Steam flow occurs only when the boiler water has reached a boiling temperature and has built up a certain amount of pressure. The system is simple and does not require a pump for circulating steam. The steam flows through the main, to riser pipes, and to the radiator. The steam condenses in the radiator and the water flows back to the boiler.

The simplest system is the one-pipe system illustrated here in which steam and condensation move through the same pipe—the two do not mix, and flow in opposite directions at the radiator valve. The piping starts high and gently slopes downhill until it reaches the boiler.

In this system only one opening to the radiator exists, with the exception of the air vent at the top of the radiator. The temperature in the room is controlled by either opening the steam valve wide or

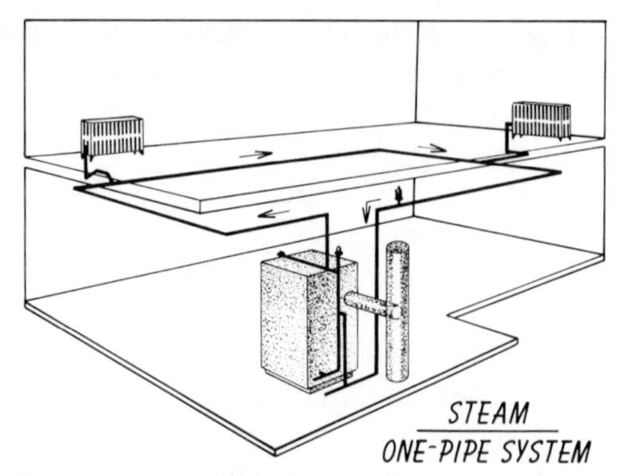

closing it completely. The system does not work satisfactorily by partially opening the radiator valve. This type of system is apt to produce a loud hammering noise as the steam entering the radiator carries along the condensation and slams the water particles to the far end of the radiator.

The two-pipe system has a separate pipe for condensation. If the steam radiator has an inlet connection at the front, and a steam trap and pipe at the other end, it is a two-pipe system. This system has better temperature control and is less noisy than the one-pipe system.

In a two-pipe steam system the steam valve can be adjusted to permit variable amounts of steam into the radiator. Hence, in mild weather the valve will be partly open for a small amount of steam, thus preventing overheating of the living area.

Steam radiators become ineffective if the radiator bottles up too much air. To prevent this, automatic air vents are provided near the top of the radiator so that air, but not steam, will be removed as fast as it accumulates in the radiator. The steam trap placed low in the radiator at the discharge end permits condensed steam, but not live steam, to leave the radiator. In either case, the fittings on a radiator should not leak steam into the surrounding air.

Radiators may consist of freestanding units below windows, partly recessed units in outside walls, completely enclosed units, freestanding convection units below windows, wall units attached to outside walls or the ceiling, or baseboard units along the outside wall.

Regardless of the shape or location of the radiator, the function of a radiation surface is to convey heat from that surface to the surrounding environment, and this can occur both by convection, as room air passes over the heated surfaces, and by radiation from the heated surface directly to the colder surfaces in the room. Most of the heat transfer occurs by convection.

When greater heat output is required for a given space, some of the following measures might be tried:

1) Force room air over the outside of the radiator by means of a fan.

2) Paint exposed radiators with an enamel-based paint of any color. However, if you want to reduce the heat output, paint the radiators with aluminum or bronze paint.

3) Place aluminum foil behind the radiator so that an air space exists between the foil and the outside wall. The foil serves as a radiation shield and reduces the heat loss behind the radiator.

New pipe installations to improve the distribution of steam will be costly and require a competent installer.

Steam leaks should be fixed. This may require repacking of radiator valves, replacement of steam traps, air vents, or pipes.

It is possible to convert an old steam system into a forced hot-water system, but this is a major undertaking and a costly one. Obtain estimates from heating firms before you begin.

Replacement or patching of old pipe insulation will help prevent steam condensation before it reaches the radiators.

HOT-WATER GRAVITY SYSTEM

This system is also almost obsolete. There is nothing essentially wrong with the system, which requires no pump for circulating the water, except that systems that cost less and respond faster have been developed.

If a gravity hot-water system has been performing satisfactorily through the years, it should continue to do so for many more years unless it becomes necessary to replace a part.

The water circulation depends upon the difference in water temperature in the two legs of the system, as well as upon the degree of slope of the supply pipe from the boiler to the radiators. *Closed* systems are provided with an expansion tank near the boiler and can operate at higher temperatures than the open system. In the *open* system the expansion tank is located above the highest radiator and the water is exposed to air.

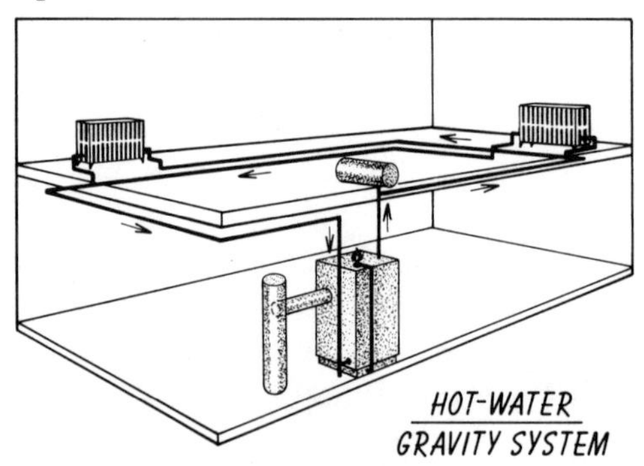

Difficulties with the system may arise if the supply water is colder than intended or if the branch piping is not properly sloped to aid in water circulation. Since a pump is not involved, the flow must be able to move freely past elbows, fittings, and piping. Pipe insulation on the supply side will reduce heat loss and aid circulation.

If this system is to be converted into a forced hot-water system, a circulating pump can be installed much more readily than in a steam system. The pipes are oversized (but this offers no problem) and the distribution of water to each radiator may have to be adjusted, but this can be done by a competent serviceman.

Since the converted forced hot-water system will probably improve the air temperature in the hard-to-heat rooms at the periphery of the house, and also operate at a lower water temperature, an improvement in fuel economy should result.

HOT-WATER FORCED SYSTEM

The large gravity hot-water system has been replaced by a forced system, in which a small circulator pump provides the pressure necessary to move water through the pipes. The piping is considerably smaller (and less expensive) than that of a gravity system. The flow of water does not depend on the heating of water in the boiler, and starts instantly when the pump is actuated.

The efficiency of heat transfer from the fuel to the radiators is higher than with a gravity system. Therefore, in spite of the electrical energy used by the pump, the overall use of energy will be less than with a gravity system.

If the system has been properly designed and installed, there is not much that needs to be done to keep it operating for many years. The pump can be wiped and lubricated, and the water level in the expansion tank inspected, but these can be done at the time the burner is inspected and serviced.

A type of boiler known as a fire-tube boiler, in which special tubes carry combustion gases from the furnace to the stack, should have these tubes inspected by a serviceman and any encrustation removed.

Two basic types of boilers are in use today: those for house-heating purposes only (which are shut down in spring), and those for house heating combined with the heating of the domestic water supply (operated on a year-round basis).

In the combined system the boiler water temperature is never allowed to become less than about 160 degrees so that hot water is always available in the kitchen and bathrooms. The advantage of this arrangement is that a separate water heater is not required.

Any heat loss from the boiler or pipes in the winter is not a true loss, since it helps to warm the house. In the summer, however, any such loss creates problems. The boiler and pipes should be inspected to make sure that the insulation is not falling off in places or perhaps was never applied to begin with.

The temperature of the water in a boiler used only for space heating varies over a wide range, depending upon the weather. In the usual control arrangement the room thermostat turns on both the burner and circulator, and turns them both off when room temperatures have reached the desired level. As the weather becomes colder the burner operates more frequently and for longer periods of time, keeping the boiler temperature higher in cold weather and lower when the heating demand is less. In modern systems the hot-water temperature can exceed 212 degrees, but steam will not be formed since the pipes will be under a slight pressure.

As mentioned earlier, if the system operates under peak load con-

ditions, in which some rooms might be overheated while others do not reach the desired temperature until after long operating periods, the fault may be either in the heat distribution or in heat loss from the living space. From an energy conservation standpoint, it will pay to reduce the heat loss first. The water supply to overheated spaces can be easily corrected by a competent serviceman.

There is not much that the homeowner can or should do with a boiler, pump, or burner except to wipe the parts, oil as per the instruction manual, and keep note of any strange sounds or happenings.

WARM-AIR HEATING SYSTEMS

Gravity Systems

The gravity warm-air heating system is, on the whole, usually to be found in older homes, especially in the Midwest. The system is simple because it consists of an oversized "stove" in the basement with a sheet-metal casing around it. The room air was removed from the living space through return-air grilles and conveyed back to the lower part of the furnace through large ducts. The heated air was delivered through large leader pipes, usually 8 inches in diameter or larger. The secret of the successful operation of a gravity system is a free-flowing system containing large ducts, relatively short, and with smooth transition joints. Where the air circulation was restricted by small ducts, sharp elbows, and long pipes, the circulating air became overheated, and the term *hot-air heating* was given to this type of restricted system.

Most of the difficulties with gravity warm-air furnace systems arise from lack of air circulation; a restriction in the return air duct is most common. Whenever the air circulation can be increased by using smooth transition pieces in place of sharp-angled bends, the warm-air temperature will be decreased and the fuel consumption will decrease in proportion.

Conversion of an existing gravity warm-air heating system to a forced-air system can be either through a series of changes in basement equipment, or through changes both in the basement and in the living space. The simpler change, which is also the less costly of the two, is to replace the old gravity furnace, remove the large leader pipes,

and remove the large return-air ducts. The new furnace will be much more compact; a new bonnet is attached, a new trunk duct or two will be installed above head level, and new branch ducts will be placed above head level. The wall stacks and register boxes will not be touched in this operation.

A slight improvement in air distribution can be made by installing air deflectors at the face of the large registers, so that the air is discharged toward the floor.

More extensive changes will create dust and confusion in the living quarters, so that they should not be approached lightly. The changes will be in the registers and grilles. The newer registers will be smaller, and they may have to be relocated. If possible, the registers or diffusers should be along the outside wall and the air should be discharged upward. As far as possible the old stacks and register boxes should be replaced, but this is not always possible. No extensive alteration of this sort should be done without obtaining estimates, or firm bids, from more than one reliable heating company.

Forced-Air Systems

The forced-air heating system is one of the most commonly used central heating systems for homes, and in many cases is combined with facilities for summer cooling. The basic elements consist of a furnace (or heat exchanger) enclosed in a casing with a centrifugal fan (called a blower) to produce a positive circulation of air. Other attachments such as an air filter, humidifier, and necessary controls are frequently incorporated in the packaged unit.

In spite of the fact that an electrically operated fan is used for circulating air, the efficiency of fuel utilization is higher for a forced-air unit than a gravity unit. This is due mainly to the more effective removal of heat from the furnace heat exchanger by the moving air. Furthermore, the casing and ducts are smaller; the branch ducts for a forced-air system seldom exceed 6 inches. If the system is for year-round air conditioning, the ducts will probably be no smaller than 6 inches, because summer cooling requires a considerably larger airflow than does winter heating. Because of the smaller ducts and the lower air temperatures in the supply ducts, the heat loss from the ducts to the surrounding air is not as great as in a gravity system, so that more of the air reaches the supply register as intended.

Air filtration is accomplished by either disposable-type air filters or by electrostatic precipitators. The electric type is highly effective in removing small dust particles down to sizes that are barely visible. Some are provided with a coarse filter for removal of lint and hair that might overload the precipitator. It is possible to expect too much from an electric filter; even if the unit removes over 99 percent of the smaller dust particles, it will not affect the dust that never passes through the unit. If clouds of tobacco smoke are being released in the living space, only that part that enters the filter unit will be removed.

Disposable air filters should be examined regularly and replaced when necessary. The air circulation through the furnace can be reduced if the air filter in the return-air duct becomes plugged with dust; under these conditions the heat transfer from the furnace decreases and the furnace efficiency also decreases. In a clean house (without long-haired dogs or new rugs that shed) the replacement of air filters should be made about once each season. Where animal fur and carpeting provide material that can clog an air filter, the replacement may have to be more frequent.

When the filters are removed you may be able to inspect the belt between the motor and the fan. (Some fans are direct drive and have no belts.) If the belt gives more than about ½ to 1 inch when pressure is applied, the belt needs tightening, usually by moving the motor base.

Normally, warm-air ducts are not insulated when the ducts pass through heated basements, since any heat loss from the ducts would be transferred to the basement air. There are several specific locations where duct insulation should be provided:

Wherever ducts (warm-air and return-air) pass through an attic space, the ducts should be well protected. Normally, a 1-inch-thick batt is used to cover the ducts. An additional 2-inch batt should be applied by draping the insulation over the ducts and allowing it to fall to the ceiling insulation. The insulation can be readily applied and the only precaution necessary is to press the edges together so that no space exists between the batts.

Wherever ducts pass through a crawl space that is exposed to the outdoor air, insulation should be provided. Any heat lost from these ducts escapes to the outdoors and is totally wasted.

For enclosed crawl spaces, a few precautions bear repetition. The earth floor in the crawl space should be completely covered with a vapor barrier, consisting of a polyethylene sheet that is carried up the sidewalls of the crawl space. The vent openings in the crawl space should be closed. The walls of the crawl space should be insulated. If these measures are taken, the warm-air ducts in the crawl space need not be insulated. Any heat loss from the ducts will serve to maintain a warmer crawl space that will prevent freezing of water pipes, and at the same time provide for warmer floors in the space above.

When the furnace is located in a garage and is partially exposed to cold air, the ducts and the bonnet should be completely protected with insulation. Better energy usage will be obtained if the furnace can be enclosed in a furnace room that is lined with fire resistant material and provided with an opening for combustion air. Find out about safety requirements from a building inspector before starting this project.

Furnaces located in a crawl space require air for combustion, and for this reason they are usually exposed to cold air. The ducts, both return and supply, should be heavily insulated with 2-inch-thick batt insulation.

Many forced warm-air installations are not properly operated for the most efficient use of fuel energy. The most desirable condition is one of almost continuous operation of the furnace blower whenever the outdoor air temperature drops below the freezing point, even though the burner itself may be operating intermittently. This condition can be brought about by setting the cut-in point of the fan switch in the furnace casing at about 110 degrees and the cut-out at about 85 degrees. This adjustment should be made by the serviceman and not by the homeowner. It is possible that with the cut-in point at this low value, the homeowner may complain of cold drafts from some registers that force air into the living area. In such cases either the airstream can be deflected away from the occupant by adjusting the vanes in the register or the cut-in setting of the fan switch can be moved upward a few degrees.

In any case the setting of the fan switch should be no higher than 150 degrees. With a setting this high the blower will operate intermittently in cool weather, and blasts of hot air from the register will flow intermittently into the living area, followed by long off-periods when the cold air from the windows settles to the lower part of the rooms.

Although with continuous air circulation the fan does run longer and the operating cost of the fan motor might be higher than with intermittent operation, the saving in fuel will more than compensate for this increase in electrical consumption. The warm-air furnace is designed to operate with a positive air circulation over the heat exchanger, and in a properly operated system the air issuing from the registers will be warm air and not hot air. There is a substantial difference between the two.

Self-Contained Heating Units

Individual through-the-wall heating units are used in warm climates and in low-cost structures or room additions. They may be operated by gravity flow or by a circulating fan. The performance of these units leaves much to be desired.

Zone Control

In large homes with two or more distinct wings, the use of zone control makes it possible to obtain more uniform temperatures in different parts of the house. Zone controls can consist of either one furnace with two or more main ducts each with its own damper control or two or more furnaces, one for each zone.

When remodeling a large old house, a small individual furnace can be used for the new addition or to handle a section of the house that is far from the old furnace.

DOMESTIC HOT-WATER SERVICE

Hot water in a centrally heated home is usually provided by a separate tank of water with its own gas or oil burner or electric heating element. The case of the heating coil immersed in a steam or hot-water boiler has been mentioned earlier. In the case of a hot-water boiler water is heated in a separate pipe loop that passes through the boiler water. The boiler water has to be maintained at a temperature of about 160 degrees or slightly higher at all times, so that during mild weather or in the summer hot water is obtained at great cost in terms of fuel energy.

The separate tank with its own heater can be either gas-fired, oil-fired, or electric. In old models the tanks were not insulated, so that the galvanized tank was exposed to the bathroom or kitchen air.

In these older installations the tank should be insulated with at least a 1-inch layer of batt-type insulation. Perhaps the tank might look better with smaller vertical strips of board-type insulation held together by adhesive tape.

Modern versions of gas-fired or electric domestic water heaters are well protected so that the surface of the heater is not hot to the touch. The only place where heat loss occurs is at the exposed hot-water pipe leading from the top of the tank to the horizontal water pipe. Any heat lost from this area does serve to heat the occupied spaces, so that in the winter it does not constitute a true heat loss. However, in the summer this loss from the water pipes makes it more difficult to cool the house and should therefore be prevented.

Any heat loss from the water pipe to the surrounding area makes it necessary to use more water, so that a waste in energy does occur. Use pipe covering on hot-water pipes.

Since sludge can accumulate in water tanks, it is good practice to drain the water from the bottom of the tank periodically. Depending on the location and the type of water, the draining process can be as infrequent as once a year or as frequent as once a month. The water should be flushed and drained until it no longer shows any discoloration.

9

Air Conditioners and the Heat Pump

Modern cooling equipment has made it possible for many people to endure the rigors of summer heat and humidity that occur in many parts of the United States. Few places in the world have the contrasts in weather that can occur yearly in large parts of this country. Following a winter of sub-zero weather, the summer can bring air temperatures up to 100 degrees and relative humidities up to very uncomfortable levels. In fact, without summer cooling, indoor relative humidities in excess of 80 and 90 percent have been recorded. These conditions bring about the growth of mold on shoes and leather goods and the severe rusting of tools and other steel objects.

Evidence exists that summer cooling can be effective in reducing the stresses of summer heat and humidity on those with heart ailments. A cooler environment not only reduces the stress on the human body at work, but also enables people to sleep more comfortably. It is most unlikely that Americans will give up summer cooling regardless of energy shortages. In some areas of the country it has become a matter more of survival than convenience. If summer cooling were to vanish overnight, it is possible that a mass exodus would occur from many areas near the Gulf of Mexico and the Southwest.

The emphasis in this chapter, therefore, is not on how to do without the air conditioner, but how to use it more efficiently. Research at the University of Illinois during the early 1930s reviewed the effectiveness of a large number of cooling methods: night-air cooling with both gravity and fan circulation, cooling with ice, cooling with city water, with mechanical compression devices using water-cooled condensers, the same techniques with air-cooled condensers, with gas-fired drying units, and so on. Eventually it was apparent that the compressor-type cooling unit was the most practical device for American homes. One significant idea that arose from this research was that summer cooling for homes was practical even in a region where the summer design temperature was 95 degrees and the outdoor relative humidity was as high as 40 percent.

In the years after the Second World War more and more people began to look upon summer cooling in many parts of the country not as a luxury but as a necessity.

Another conclusion that emerged from the early research was that a critical factor in summer cooling was the reduction of heat gain in a house from the outside. It is foolish to permit the sun's heat to pour into a house and then use expensive cooling equipment to remove it mechanically. Therefore, many of the items discussed in the chapter on Solar Heat Control should be considered with the discussion in this chapter.

THE COMPRESSOR-CONDENSER UNIT

The modern cooling unit is a relatively simple device that has become a low cost mass-produced item. Four separate devices make up the most common type of compressor-condenser unit. In the *compressor* a gas (called a refrigerant) is compressed. The compressor works like a giant tire pump—the gas leaves the compressor at high pressure and is quite hot. This hot gas must be cooled before it can do any cooling, and this takes place in a heat-transfer device called a *condenser,* which looks like an automobile radiator. As it cools in the condenser, the refrigerant gas changes to a liquid while still at high pressure. The liquid then passes to an *expansion valve,* which permits the liquid to expand into an *evaporator* or cooling coil. The expansion process cools the coil and it can now pick up heat from air that is moved across it. In doing so the refrigerant gas is warmed and then recycled in the compressor. The entire refrigerant system is tightly sealed. In the usual residential cooling unit the compressor and condenser are in one package located outdoors, and the expansion valve and evaporator are placed indoors.

In forced-air furnace systems used for both winter heating and summer cooling, the evaporator is usually located above the furnace casing in the form of an A-shaped coil. The furnace blower moves warm-room air over the A coil, where it is cooled and moisture condenses out on the coil. This cooled air is then distributed through the same ducts, registers, and diffusers used to deliver heat to the rooms. The return air from the rooms is collected at return-air grilles and then moved through air ducts to the filter and furnace blower. The only difference in the air-handling equipment for a combination heating-cooling system is that the airflow requirements for summer cooling are considerably greater than those for winter heating, so that the furnace blower and ducts must be made larger for summer operation.

For optimum effect cool air should enter the room near the ceiling, so it can settle slowly toward the floor, cooling the room without drafts. Also, the best place to deliver heated air is at floor level along the outside wall. Obviously, if one system is to serve both purposes, a compromise must be made. In those parts of the country where cooling is the more important function, ceiling or high sidewall outlets are used. Where heating is the major concern, the air is delivered to baseboard or floor outlets.

From the many hundreds of tests made on various air-distribution devices, the two types shown, located on the exposed walls of the house, were found to be the best compromise for year-round air conditioning in areas where heating is a major concern.

The compressor and condenser units, if located outside the house, will work best in shade. All things being equal, a location on the north side of the house is preferred in most areas of the United States. If this is not possible, a fence or a piece of shrubbery can provide the necessary shade.

Since air must flow freely to the condenser, grass and weeds that might interfere with airflow must be kept away. Grass clippings, dirt, and so on should be brushed or hosed off the condenser coils.

For protection against snow, sleet, ice, birds, and leaves, a canvas cover over the outdoor unit is a good idea when it is not in operation. Plastic is not suitable because it holds moisture inside the unit and does not permit evaporation.

The indoor unit will condense many gallons of water on a humid day, and this requires a drain line. If a hose connection is made from the A coil to the drain, it is possible that the hose will accumulate algae and block the flow. Before the cooling unit is placed into operation, the drain line should be flushed out.

Make sure that a clean air filter is in place in the furnace casing. Sometimes filters are removed and not replaced. Lint and dust that are not removed by an air filter can collect on the underside of the evaporator coil and seriously reduce its cooling capacity. Also, the coils are difficult to clean once this happens.

The evaporator coil is designed for some specified rate of airflow. For this reason you should not close too many of the registers of the return grilles in the house. As an approximation, not more than about one-quarter of the registers should be closed. Restricted airflow can cause frost to form on the evaporator to such an extent that ice blocks the flow of air. The remedy is to stop the unit, allow the ice to melt, clean the air filter, and open some of the registers.

In multistoried and split-level houses the upper stories may be warmer than desired and the lower stories too cool. The damper adjustments found desirable for winter heating will almost certainly have to be changed for summer cooling. Be sure to mark the position of the dampers with W and S to show the proper settings. In the majority of cases most of the airflow to basement rooms can be shut off in the summer, so that the air is diverted to the upper stories.

ATTIC LOCATIONS OF EVAPORATORS AND FANS

For hot-water and steam-heating systems a separate air-distribution system must be installed for cooling. One common method is to place the system in the attic and discharge the cooled air through horizontal ducts to ceiling registers or diffusers. The return-air grille must also be located in the ceiling. A drain line must be provided for the moisture condensation from the evaporator, and this line must discharge on the outside of the house. Any leakage could damage the ceiling, so tight connections must be made. A pan connected to a separate drain line is often installed under the evaporator unit.

Since all attic duct work is located in the hottest part of the house in the summer, with temperatures ranging from 120 to 140 degrees on hot sunny days, the ducts must be heavily insulated. It is common practice to use a 1-inch-thick layer of insulation inside the duct. Additional insulation consisting of 2-inch-thick batts of mineral wool draped over the duct until it meets the ceiling insulation will be a

worthwhile investment. Remember, the cooling of air is one of the most expensive processes in the house.

The attic space must be ventilated to keep it at reasonable temperatures during hot summer days. Openings near the top of the roof, as well as near the eaves, permit a natural ventilation that will help considerably.

FAN-COIL UNITS FOR WATER-HEATING AND COOLING SYSTEMS

Summer cooling and winter heating can be combined in a one-room unit. In this type of unit, hot water is pumped through the radiator coil in winter, and the fan in the cabinet blows air through the coil and discharges warm air into the rooms. The usual location of the unit is below the windows, but it may also be recessed in the stud space in the wall. In the summer chilled water is piped through the radiator coil; the water is obtained from a water chiller that replaces the evaporator coil described earlier. A pump is required to force the water through the coil. This type of unit is inclined to be noisier than the remote attic unit, and is most often found in multifamily dwellings. The effectivness of the unit will be maintained by periodically replacing the air filter.

ROOM AIR CONDITIONERS

In the room air conditioner the four parts of the compressor-condenser system are in one package that is mounted on the window ledge or in a sleeve through the wall. The compressor-condenser portion extends outdoors, and the evaporator and circulating fan face indoors. These units are noisier than central units.

All things being equal, a room air conditioner should be installed in a wall facing north or east.

REFRIGERATION TONNAGE

The cooling capacity of equipment, just as in heating, is expressed in terms of Btu per hour. In the refrigeration industry 12,000 Btuh is defined as *1 ton cooling capacity.* This practical unit defines the cooling capacity of 1 ton of ice melting in a 24-hour period. Room units are available with capacities ranging from as small as 5,000 Btuh to as large as 35,000 Btuh, but units of about 12,000 Btuh are the most common.

Central cooling units, on the other hand, can be obtained in almost any size from 12,000 Btuh and up. The common size for residential service is a 30,000 to 36,000 Btu capacity unit.

Cooling units perform best when their capacity is somewhat less than necessary for extreme conditions. That is, oversize cooling units are actually detrimental to good humidity control.

The oversized unit will operate intermittently, and during the on-cycle the relative humidity indoors may drop sharply, perhaps as much as 20 percent in a 30- to 40-minute period. However, when the compressor stops, the moisture clinging to the evaporator coil is quickly reevaporated and the relative humidity will rise even more quickly than it drops. These alternating periods of dryness and clamminess result in a very high degree of discomfort.

The slightly undersized compressor operates almost continuously on a hot day and the humidity decreases somewhat more slowly than with the large unit. However, the humidity stays down and continues to fall as the hours pass and the unit keeps running. This humidity control problem is not experienced in the arid West, but does affect large areas of the remainder of the country.

In central cooling units there are two possible methods of operation.

1) In the frequently recommended method the compressor is turned on and off by the room thermostat and the circulating blower is operated *continuously*. The advantage claimed is that sustained air motion is more pleasing to the occupant and provides more even temperature distribution.

2) The alternate method of operation is to have both compressor and fan operated by the room thermostat. The energy saving is in the blower operation, which can be substantial, especially on milder days. The advantage of this method is that the increase in humidity after the compressor has stopped is not as great as with the first method. The fact that air movement is not sustained is not a serious limitation.

Set the blower to turn on and off with the compressor. This is usually controlled by a switch on the cooling thermostat. It should be set to "automatic" or "intermittent."

Commercial installations and automobile air conditioners often control humidity by the "reheat" method, and this has been suggested for residential cooling as well. In this method the cooling unit operates until the proper humidity is obtained, and then the air is reheated until the desired temperature is reached. Reheat systems should not be used in residential service where humidity control is not that important.

It is possible that the energy consumption of two small capacity units may be less than one large unit, but this needs engineering study. However, it may be desirable to divide the cooling load of a house into two unequal parts. If, for instance, a large house requires 72,000 Btuh, the load could be split into two areas requiring 24,000 and 48,000 Btuh. The unit controlling the smaller area could be set to operate at 77 degrees and the larger unit at 79 degrees. On a warm day the small unit would start first and the humidity would begin to decrease. (The humidity is relatively uniform throughout the house, even if the temperature is not.) The cooling capacity of the unit would be satisfactory on mild days, but would be unable to control room temperatures on warm days. After a period of time the larger unit would begin to operate, but only intermittently as needed to control the temperature, while the small unit would operate almost continuously. The advantage would be that humidity control would be far better than with a single unit turning on and off at 78 degrees. Since lower humidity does result in less discomfort at higher room temperatures in the summer, there may be an energy saving that would offset the lower efficiency of the smaller units. This would have to be determined by experiment and analysis.

ABSORPTION EQUIPMENT

This is one type of central cooling system that depends on heat from gas or steam rather than electricity. It essentially duplicates the operation of the compressor-condenser unit. Chilled water is produced and this is supplied to cooling coils. There are relatively few of these units in houses at present, in spite of the small number of moving parts and the quietness of operation. Higher initial cost has been the main reason for this.

EVAPORATIVE COOLING

In the desert areas of this country advantage can be taken of the fact that when water is evaporated, heat is removed from the air. A typical evaporative cooling unit consists of a cabinet containing a moist pad or filter and a fan. Dry outdoor air is drawn through the pad, which is constantly moistened. The water evaporates, the air is cooled and humidified, and this cooler, moister air is distributed throughout the house. The indoor relative humidity increases somewhat, but in dry regions the increase can be tolerated. It is possible to take 90 degree air at 10 percent relative humidity and cool it to 65 degrees, which, when warmed to 78 degrees would have an acceptable relative humidity of about 40 percent. This method of cooling works only in the dry areas of the Southwest and in certain portions of the Northwest, east of the Cascade Mountains. It should not be tried in humid areas.

NIGHT-AIR COOLING

In the 1930s an extensive series of investigations was made of the so-called "poor man's air-conditioning system," the name given to night-air cooling by fan circulation. In this system the windows and doors were opened after sundown (about 7:00 P.M.) and large quantities of outdoor air were drawn into the house by ¼- to ½-horsepower fans. The air circulation provided by the fans, commonly located in the attic and drawing air through the windows, varied from 15 to 30 air changes per hour. The higher rate means that the air in the house is being changed every 2 minutes. Even on a calm warm night the curtains would be blown inward and a noticeable air movement would exist wherever the windows were open. The size of the fan was commonly limited to the size of the window opening in the attic, but in the Southwest large fans with diameters of several feet were installed at one time.

The fan was operated continuously until the indoor air had dropped to some level governed by a room thermostat. In drier climates, where nighttime temperatures were frequently 30 or more degrees below maximum daytime temperatures, the flushing action of the fan eliminated a large amount of accumulated heat. In humid climates the disadvantage of the system was that humid night air was brought into the house, and when the windows were closed in the early morning, the humidity was very high.

Another limitation was that this method was not entirely suitable for areas where lower-story windows had to be locked at night. If the upper-story windows alone were opened, the cooling effect was reduced. Furthermore, the system could not be employed in areas where the night air was dusty.

Nevertheless, in spite of its obvious defects, it has a place in many areas, especially where nights are cool and the air reasonably dry. The system will not work efficiently when the structure being cooled is not well insulated. This is because as soon as the windows are closed, at about 7:00 A.M., the house starts warming up rapidly, and by noon indoor conditions are unbearably warm. In a well-insulated and solar-controlled structure, however, the heating rate from the sun and the outdoor air can be as low as ¾ degree per hour. If the house has been cooled at night to 65 degrees, for example, it might take all day for it to warm up to 78 degrees, and by that time the next cooling cycle could be started.

In many installations an attic fan for night-air cooling has been used as a supplement to mechanical cooling. The attic fan in this case is fairly small, perhaps with a ⅓- or ⅙-horsepower motor, and the method of operation not quite as demanding as the system described above. As an energy-saving measure, this arrangement makes sense, provided that the owner is willing to experiment with the device until the best operating conditions are determined for various types of weather and for his particular house. For those afflicted with ragweed hay fever this method is not suitable during July and August, and these individuals should live in completely closed houses that are mechanically cooled and provided with filtered air.

In summary, the basic rules for a successful night-air cooling operation are as follows:

a) Open the windows after sundown or whenever the outdoor air feels as cool as the indoor air.
b) In a two-story house, close the first-story windows if safety requires that the house be locked, but do so just before retiring.
c) Operate the cooling fan *all night long* and get rid of the accumulated heat in the house. A thermostat control can shut the fan when the indoor temperature reaches a satisfactory level such as 65 degrees.
d) In the morning *close the windows early,* preferably before 8:00 A.M., even though the outdoor air feels fresh and cool. The secret of night-air cooling is to keep as much heat out as possible, and to use the insulating capacity of the house

to keep the indoor air at a comfortable level. In a well-insulated and solar-controlled house the indoor air temperature will obviously not rise as fast as the outdoor air temperature.

e) In a long, narrow mobile home, a ventilating fan, big enough to fill a window that is installed at one end of the home, can draw air through the entire structure if the windows at the opposite end are opened. Normally, mobile homes are not well insulated, so that the cooling effect will be dissipated by early afternoon. In spite of this limitation, a fan can offer low-cost cooling during night hours.

In the face of the energy crisis many "new" ideas will be introduced to assist in cooling homes. Many will be akin to reinventing the wheel—they have already been tried in years past and found to be wanting. A few of these are described below.

ROOF PONDS

Tests have been reported where flat roofs were covered with a 1-inch-deep pool of water exposed to the sun. The water is heated, and some evaporation takes place. The heat gain through the roof was reduced by more than 65 percent this way. Unfortunately, the method is not practical for many homes. The roof must be able to hold water, and the cost of constructing such a roof is considerably higher than the cost of the usual roof. Also, the water becomes polluted from dust, leaves, birds, and algae; the pool has to be drained before cold weather sets in; and a water replenishment system must be installed. The savings are really not worth the expense and trouble.

ROOF SPRAYS

Tests have shown that the heat gain through roofs can be reduced about 75 percent by means of a fine spray of water on roofs exposed to the sun. The system works best in a dry climate, and where a plentiful supply of water is available at low cost.

The roof construction must be able to withstand the constant wetting (the life of the roof is often significantly shortened), and provisions must be made to handle the run-off water. The cost of water can be prohibitive where city water is the only source of supply.

CAVES, UNDERGROUND PIPES, ETC.

Installations do exist where naturally cool air has been used to air-condition buildings. For example, air has been piped from natural caverns that maintain a year-round temperature of about 50 degrees. If 50 degree air at 100 percent relative humidity is pumped into a structure and allowed to heat to 75 degrees, the indoor humidity would be about 40 percent, which should be acceptable. If you happen to live over a cave, it is something to think about.

Air has also been piped from deep shaded ravines and pumped into buildings, another unusual possibility in certain areas of the country.

It has been proposed that large drainage pipes be buried in the ground and that outdoor air be drawn through them before it is pumped into a building. Even assuming that 24-inch drainpipes several hundred feet long are buried about four feet below ground, the heat transfer into the ground would be slow. It is true that cool air might be obtained for a few minutes, but the surface is not large enough for practical cooling.

THE HEAT PUMP

The heat pump is a practical, commercially developed machine that makes use of the principles of refrigeration to provide both a winter heating system and a summer cooling system. The principle is referred to as the reversed-refrigeration cycle and was developed in the nineteenth century.

In a refrigerator the heat generated in the compression process, together with the heat picked up by the evaporator coil, is discharged through the condenser coil. In the ordinary kitchen refrigerator, for example, the condenser coil throws off the heat from the compressor and the cooling coil in the food compartment. In this case the heat is passed on to the kitchen itself.

In residential use a heat pump is essentially a compressor-cycle air-conditioning apparatus in which the direction of flow of the refrigerant can be reversed. In the summer the condenser is located outside the house to discharge heat and the evaporator coil inside the house to absorb heat. In the winter the same condenser becomes the evaporator and removes heat from the outside air. The coil inside the house becomes the condenser and discharges the heat generated by the compressor *and* the heat removed from the outside air by the evaporator into the living area.

When the outdoor air temperature is mild, say 50 degrees, a large amount of heat is picked up by the evaporator, so that the heat output at the condenser can be as much as four times the heat equivalent of the power required for the compressor. This is equivalent to getting four Btu for the price of one, since three of these heat units were picked up in the evaporator from the outside air and only one was paid for with electrical energy.

If the outdoor temperature drops to the same level as the evaporator temperature, about twenty degrees, then no heat absorption will take place at the evaporator. The efficiency of the system is therefore reduced, and there is no point in operating the compressor unit, so a supplementary electrical heater is automatically turned on.

The heat pump was originally popular in milder climates, where the cost of electricity was low, and where alternative sources of energy were not plentiful. The use of heat pumps has now spread to colder climates because many winter days are not all that severe, and the heat pump offers a means of using less energy than an ordinary electrical heater.

The earlier heat pumps experienced difficulties, some of them mechanical and others related to the working of the control devices. One problem that proved annoying was the rapid frosting of the evaporator coil under conditions of cool, damp outdoor air—a defrosting mechanism had to be used so that accumulated ice would fall off the coils before normal operation was started. The severity of the service requirement can be understood when you realize that the unit may have to operate in summer weather of over 100 degrees as well as in winter weather that may drop below zero. This type of equipment, operating 4 months of the year only as a cooling system, can be expected to last 10 to 15 years, but operated year round as a heating *and* cooling system, the mechanical parts cannot reasonably be expected to last the same length of time. Fortunately, the problems of lubrication, sturdy control equipment, and reliable valves have been largely overcome in recent years.

The heat pump system has many advantages, although not in every situation. The initial cost of the heat pump is higher than an electrical or fuel-fired system, but comparable to a conventional heating system plus central air conditioning. It will also probably use less energy than conventional systems.

The best advice that can be given to any homeowner considering the installation of a pump is to find a reliable heating company in the area that has had some prior experience with the equipment. Talk to someone who had a heat pump installed. Find out what happens when the unit becomes inoperative, when the power supply fails, or when a brownout occurs and reduced voltage is supplied. Find out how the unit can be shut off and how straight electrical heating can be turned on. Find out what service charges will be and what service guarantees apply. Then compare the pump to more conventional systems.

10

Solar Heat Control

Solar energy can work for us or against us. By properly designing our houses, solar energy can be used in the winter to reduce heating bills without the need for specially designed solar collectors and heat storage devices. At the same time it is possible to use shades of various types to keep out solar energy in the summer.

WINTER SOLAR HEAT GAIN

The path of the sun in the winter sky depends upon the latitude of the region. The 40-degree latitude line cuts across many populated cities in the Midwest and the East. At this latitude the sun's path on the shortest day of the year is about 30 degrees above the horizon. It is apparent that the sun's radiation strikes the south wall of the house almost directly. The solar intensity on the south wall is greater in the winter than it is in the summer when the sun is higher in the sky and the solar radiation strikes the south wall at a sharper angle.

It is also important to note that there are fewer daylight hours in the winter and that the sun does not shine on east or west walls directly but always obliquely. This means that in the winter the east and west walls do not receive much solar energy, and that only for about two or three hours each.

With these facts in mind it is apparent that if solar heat is to be used to heat a home in the winter, the south wall is the key.

By means of a properly designed roof overhang it is possible to prevent solar energy from entering the house during the summer, but to permit it to be absorbed in the winter. The sunlight that enters through windows is absorbed by the floors and furnishings in the house, but the longer wave lengths (infrared radiation corresponding to temperatures near the 100 degree level) are trapped indoors, and eventually warm the air in the house. The rooms on the south side of the house will be warmed, while those on the north side will be unaffected, unless the air on the warm side can be distributed to the cooler parts of the house.

As part of solar control, the role of trees and bushes should be considered. Tree shade on the east, south, and west sides of the house would be desirable in the summer, but not in the winter. In the temperate zone an ideal arrangement can be achieved by planting trees that drop their leaves in fall and allow solar radiation to reach the walls.

When the winter sun shines on the south wall, or on the roof, the radiation warms the surface and to that extent reduces the heat loss from the house. When outdoor air temperatures are about zero degrees, the surface temperature of a south wall can still be as high as 60 degrees. Theoretically, a dark surface would be better for absorbing solar radiation in the winter, but would not be desirable during the summer when it is necessary to reflect solar radiation. The choice of color might depend upon the geographical location of the house, and the relative importance of heating and cooling for the region. If the wall or ceiling is well insulated, the color of the wall and roof makes little difference.

SUMMER SOLAR HEAT GAIN

Sun control in the summer is of tremendous importance. Not only does a lack of control result in the need for large and costly refrigeration equipment, but also in high operating costs. At 40 degrees north latitude, a rule-of-thumb used by some utility companies is that summer cooling costs for 3 months will be about the same as the heating costs for the remainder of the year. On a monthly basis, therefore, summer cooling can cost from two to four times the average monthly winter heating costs.

In far too many cases little attention is paid to keeping the sun out of the house, and oversized cooling equipment is installed to overcome the heat produced by the sun. This is poor planning and poor operation. In most parts of this country the peak load for the electrical utility companies occurs in the summer as a result of air cooling, and a large part of this peak load is imposed by residential cooling, both with window and central air-conditioning units.

The path of the summer sun on the longest day of the year, at 40 degrees north latitude, places the position of the sun at almost 75 degrees above the horizon. (As one moves south the noonday sun is almost directly overhead, 90 degrees above the horizon, whereas farther north the angle is smaller.) Furthermore, the summer day is longer and the east wall receives solar radiation for over 6 hours in the morning. The same is true of the west wall in the afternoon and early evening.

The intensity of solar radiation in the summer shows the following maximum values on a clear day.

Position of Surface	*Time of Maximum Intensity*	*Heat Intensity at That Hour (on 1 square foot)*
Flat roof	Noon	290 Btuh
East surface	8:00 A.M.	220
Southeast surface	9:00 A.M.	175
South surface	Noon	90
Southwest surface	3:00 P.M.	175
West surface	4:00 P.M.	220

It is apparent that the east and west surfaces and the roof receive large amounts of solar radiation. The roof surface is exposed to radiation longest of all.

ROOF SOLAR CONTROL

The worst possible type of roof would be a thin layer of metal between the occupant and the sun—the lower surface of the roof would be a broiler whose temperature could be in excess of 160 degrees.

As construction material is added to this roof, the heat radiation from the sun is absorbed by the material, so that the heat flow through the roof is delayed. However, if insulating material is not provided in the roof construction, the heat flow can be rapid and the undersurface of the roof may still reach broiling temperatures. (Fishing cabins in the north woods are built without any insulation and the roof radiation can be intense and almost unbearable in the heat of a summer day.)

If sufficient insulation is applied to the exposed roof, the heat flow can be slowed down and the temperature of the inside roof surface can be reduced. However, the heat stored in the roof structure will show up many hours later, perhaps when it is time to go to bed.

The attic space serves a useful purpose in reducing the effect of roof heat gain. The roof in this case acts like an umbrella, and the attic air temperature is warm, but never as hot as the roof surface. The proper solar control requires that the attic space be vented so that outdoor air, at perhaps 90 degrees, can enter the space and replace the 120 to 140 degree air that has been trapped there. This may require vent openings in the soffit, or overhang, as well as large openings at some high point in the attic space. This can consist of attic windows, attic louvers, ridge vents, or even openings at the top of the roof.

In some roof and attic constructions, openings for natural venting may be difficult to install. Power-operated vent fans are available, but since these are energy consumers, it becomes questionable whether they should be used. If the forced venting will reduce the attic temperature sufficiently to reduce the load on the cooling unit, perhaps the vent fan can be justified.

In a normal peaked attic for which natural venting was possible, if the cost of a vent fan were invested instead in additional insulation, the heat gain would be reduced at least as much as by the use of the fan, and the insulation has no operating or maintenance costs.

The use of ceiling insulation below a vented attic space is the best construction method. By installing 6 inches or more of insulation in the ceiling, exposed to the attic space, the heat flow is made negligible and the ceiling surface temperature approaches room temperature.

In some mobile-home parks in the hot Southwest, a complete roof structure is erected on a pole frame over the mobile home, the carport, and outside living space.

SOLAR CONTROL ON WINDOWS

Windows that face east or west, as well as southeast or southwest, should have solar control devices. The most effective method is to intercept the radiation *before* it strikes the windows. This can be done by means of tree shading, outside shutters, awnings, outside venetian blinds, and a special type of window screen known as the shade screen. In the shade screen the wires are replaced by thin narrow strips of metal that are held in place at a fixed angle and prevent solar radiation from passing through until the sun is quite low in the sky.

The second line of defense against solar radiation is *at* the window surface. Several manufacturers produce a "sun control film" made of plastic with an extremely thin reflective metallic coating that permits enough light to penetrate for clear vision but reflects most of the sun's heat. It appears mirrorlike from the outside during the day and from the inside at night. It may be applied to existing glass by the homeowner or professionally, although it is relatively expensive. Factory-coated plate glass with similar properties is also available and is sometimes used in office buildings. Reflective glass surfaces have the disadvantage of also reflecting heat during the winter.

The final line of defense is the use of shading devices on the *inside* of the window. These include window shades, venetian blinds, and full-length draperies.

In general, those shading devices on the outside of the window intercept the radiation, and when they become heated, convection currents carry the heat to the outdoors. The reflecting films that are pasted on the window, as well as heat absorbing glass, reflect part of the radiation to the outdoors and absorb part, with the remainder passing into the room. In the process the glass becomes heated and some of this heat is passed into the room. Internal shading devices such as shades, blinds, and draperies are effective if they are light-colored to reflect the sun and if they cover the entire window.

Apply window-shading devices on east windows, then west windows, and finally the south. The east windows have high priority because the heat that enters the house in the morning may activate the house cooling system early in the day.

Awnings have an effectiveness similar to that of louvered screens.

11

Home Appliances

Americans are probably more dependent on various home appliances than people of any other nation in the world. As a consequence our appliances are major users of energy.

Selecting an energy source for major appliances is a complicated problem. There is no best source of energy for heat-producing appliances. In some instances electricity may be more efficient than gas and in some instances the reverse. Though electricity might be very efficient at the *point of utilization,* the home, the process of generating electricity for the community as a whole may not be a very efficient process, so that the amount of energy used in the total system to produce a unit of heat in the home may be considerably more than for a less efficient fuel. If, however, the electricity is produced by a relatively plentiful source of energy like coal, or hydroelectric power, or nuclear energy, and the alternative home-heating fuel is gas, which is in limited supply, the lesser overall efficiency of generating electricity may be preferable. If electricity is generated with oil, it is indeed folly to use electricity instead of gas, which, though producing heat at a lower level of efficiency in your home, is more efficient on a system-wide basis.

It is clear that in the future our choice of an energy source in the home may be severely restricted by the social cost of making that energy available.

ROOM AIR CONDITIONERS

The appliance that consumes the most power in the American household is the room air conditioner.

Since room air conditioners are usually just plugged into existing wiring, be sure that this wiring is adequate for the current load the air conditioner will draw. The nameplate on the air conditioner should indicate the number of amperes that the machine will demand. If the amperes on the nameplate exceed 15, the house should have special wiring in order to carry the power safely. It is desirable that all air conditioners be on a separate circuit, with no other appliances or lights on the same line. Most circuits in houses are designed to carry a maximum of 15, or sometimes 20, amperes without being overloaded.

Nearly any other appliance on the same circuit, operated at the same time as the air conditioner, will overload the circuit, using more energy and causing an electrical hazard. The energy is lost because an electrical wire becomes hotter as the current flow increases. The heat generated in the wire increases with the square of the current, which means that doubling the current flow in the wire quadruples the heat produced. The heat produced by this energy is either completely lost or comes into the house and must in turn be removed by the air conditioner.

Preferably, room air conditioners should not be operated on extension cords. If it is impossible to rewire to operate the air conditioner on its own outlet and circuit and an extension is needed, a special heavy wire extension cord, no longer than is absolutely necessary, should be used.

Location is important for room air conditioners. If possible, the unit should be located in a north or east window, or at some point where it is shaded during the time it operates. Since the air conditioner must transfer heat from the house to the outdoor air, using a condenser similar to the radiator in a car, the heat transfer is much more efficient if the outdoor air temperature is lower. Direct sunlight on the outdoor portion of a room air conditioner can reduce its efficiency by 10 percent or more.

In shading the air conditioner, you must be careful not to restrict the airflow around the unit, since large quantities of air are forced through the condenser by the cooling fan.

The built-in efficiency of an air conditioner is an important factor in

reducing power consumption. The nameplate of each air conditioner should indicate the Btu cooling capacity and the power requirement, either in terms of watts or volts and amperes. Multiplying volts times amperes gives watts for purposes of comparison. In selecting an air conditioner you should divide the number of watts required into the number of Btu of cooling produced. This will give you a rough measure of the unit's efficiency.

The average room air conditioner has a rating of about 7 Btu per watt. The range is from 4 to 10 or 12. If all 35 million room air conditioners in the United States were of the high efficiency type, we would save 26 billion kilowatt-hours per year.

A room air conditioner operates in exactly the same way as a central air-conditioning system. The only difference is that all the components—compressor, condenser, evaporator coil—and fan, are combined into one cabinet, which can be easily installed in a window opening or a sleeve built into a building wall.

The same operating procedures for a central air-conditioning system apply to the use of room air conditioners. In addition, the cooling effect of the room air conditioner may well be increased by the use of fans to move the cool air around the room, since there is only one cool air outlet instead of the several outlets in a central system.

A room air conditioner is less efficient when operated on low fan speed. Though the noise of the fan and air movement is reduced, the air produced will be colder than if the fan were on high, perhaps causing uncomfortable drafts, and the heat transfer is less efficient.

DOMESTIC WATER HEATERS

Selecting an energy source for domestic water heating is primarily a problem of availability. Electric water heaters are, of course, 100 percent efficient since the heating element is inserted directly into the water to be heated. Gas and oil-fired water heaters have heat loss up the vent stack as the products of combustion are vented to the chimney.

Another major factor in water heater selection is the *recovery rate,* or the number of gallons of water that can be heated per hour as the water is used. This can be very important if there is a specific washday in the household, where four or five cycles of the automatic clothes washer may occur during one working day. Electric water heaters, even the quick recovery type, have a significantly slower recovery than gas- or oil-fired heaters. Gas-fired water heaters are available with two-stage burners, so that if a large quantity of water is drawn, a larger section of the burner will ignite, increasing the recovery rate. Again, these larger units are less efficient. The oil-fired water heater has so much heat input that recovery is almost instantaneous. As long as only one hot-water faucet is open, the oil-fired water heater can provide a continuous stream of hot water 24 hours a day.

Selection may also be based upon space available, since the lower recovery rate heaters require larger storage tanks. Gas- or oil-fired heaters will also require a vent or flue.

The critical factor in reducing the energy consumption of water heaters is the reduction of the actual amount of hot water used.

Be sure that there are no leaks in faucets or appliances attached to the hot-water line.

Insulate the water lines from the water heater to the points of use. This will reduce the amount of water that has to be drawn off before the water becomes hot.

Smaller size lines may be used for sinks and so on to reduce the total volume of water contained in the pipe between the heater and the water-using appliance.

Set the water temperature control on the heater at no more than 140 degrees. Automatic dishwashers will compensate because they have heating elements that maintain or increase the water temperature, and higher temperatures are not generally required for most household activities. The lower the water temperature the less heat is lost from the water heater and the water lines.

It is more efficient to heat small quantities of water for instant beverages and so forth on the stove, rather than by running water from the faucet until it becomes hot enough for such use.

A tempering tank, an uninsulated steel tank installed in the feed line leading to the water heater, will use the heat in the house to warm the incoming water to room temperature. Using such a tank, there is often a 20 to 30 degree increase in water temperature before it even reaches the heating element, and if the water is warmer going into the heater, less energy will be used and the recovery rate will be faster. In the summer the heat gained by the water is free, but in the winter this heat will come from the central heating system. Condensation may occur on the tank during the summer if the space in which it is located is not dehumidified, and provisions must be made to drain away this moisture. The tank should be at least as large as the capacity of the heater.

CLOTHES WASHERS

Although the clothes washer for the average family uses only about 100 kilowatt-hours of power every year, about $3 worth of energy, it can waste much more energy because it uses vast quantities of hot water. The machine should have cold and warm water wash and rinse cycles, and be able to be adjusted to use less water for small loads. A full cycle on a top-loading washer takes about 35 gallons of water, most of it hot. This water must be pumped, treated, heated, and then processed as sewage, all of which uses large amounts of energy.

Always wash full loads. Reduce the water setting if the load is not full.

Use a warm or cold water wash and rinse cycle where possible. Remember that the bacterial count is higher when clothing is washed in cold water. Underclothing, handkerchiefs, sickroom linen, and so on should still be washed in hot water.

Wash fabrics only when they are dirty, not just because they have been used slightly. Most clothing does not wear out; it is washed to death.

CLOTHES DRYERS

Both electric and gas-fired dryers using an electric ignition system use approximately the same total amount of energy per year in drying clothes. However, a gas-fired dryer that has a pilot light uses almost as much gas to operate the pilot light as it does to dry the clothes. The ultimate in energy conservation is, of course, hanging the clothes out to dry rather than using the dryer. However, this involves the handling of heavy wet materials in all kinds of weather; it is no longer practical in many urban environments, and it is unlikely that it will become a widespread practice.

Consider removing clothes from the dryer when they are nearly dry and allowing them to finish drying while hanging on a line. In terms of energy consumption the last bit of water is much more expensive to remove than the water at the beginning of the cycle.

If the laundry is removed somewhat damp from the dryer, the pieces can be ironed immediately, saving the cost of complete drying and then redampening for ironing.

The venting of an electric dryer into the house in cold climates conserves heat and adds humidity, but it should be discontinued if it causes condensation on the inside of double-glass windowpanes. In order to avoid the release of lint into the room, a lint trap should be installed at the point where the air is discharged from the dryer. An old nylon stocking held in place with a large rubber band is a simple but fairly efficient filter. Be sure that the dryer vents to the outdoors in mild or warm weather.

Gas dryers must be always vented to the outdoors because the products of combustion should not be mixed with room air.

COOKING APPLIANCES

The choice of cooking appliances is one in which a great deal of seemingly conflicting information exists, and many invalid comparisons are made. This is true both in choosing a fuel for cooking and in choosing the type of burner or cooking surface.

The standard test for the efficiency of a burner is the ability of that burner to transfer heat to an aluminum block the same diameter as the burner. This is supposedly equivalent to heating a flat-bottomed pan containing water or food—but in an electronic range the heat is transferred directly to the food and metal cannot be used, which means that no direct comparison can be made. A similar problem exists with the new induction ranges—stainless steel is the preferred type of pan and aluminum is very inefficient. On the smooth-surface cooking units now becoming popular, heat is transmitted by conduction, which requires very close contact between the bottom of the pan and the cooking surface. Special glass-ceramic cooking utensils with ground and polished bottoms are recommended, because as metal pans get hot, they distort and the bottom is anything but flat. Great differences in efficiency result. In the following paragraphs we will try to assess the efficiency and the energy usage of various cooking appliances and discuss how these ratings were obtained and how they apply.

Electric Range

The efficiency of a conventional electric range is significantly affected by the size and type of cooking utensil.

A flat-bottomed aluminum or copper pan is the most efficient, and should be of a size that completely covers the heating coil or that portion of the coil that is being used. (A pan with a slightly concave bottom becomes flat as it heats.) Under these circumstances the large range burner has an efficiency of about 75 to 80 percent. The small burner has an efficiency of about 50 to 55 percent.

A conventional electric range doing normal cooking for a family of four will use about 132 kilowatt-hours or 450,000 Btu per month, according to studies by the U.S. Department of Agriculture.

The smooth-top range burner with an automatic temperature control and specially made glass-ceramic cooking utensils also has an efficiency of about 75 percent. But, if an aluminum pan is used instead of the glass-ceramic container, the efficiency drops to 36 percent.

Use flat-bottomed pans of the proper size to cover the portion of the burner used.

Use only the specially made cooking utensils on smooth-top ranges or hot plates.

Turn off the burners before the food is done, and use the residual heat in the burner to complete the cooking.

Gas Ranges

Gas ranges produce their heat with a flame, and do not depend on conduction from the burner to the pan, so the use of flat-bottomed pans is not critical. Pan size is very important, however. Heat produced by a flame that goes up the side of a pan is not as well utilized as that produced under the pan. The average efficiency of a surface cooking unit on a gas range is about 50 percent.

The gas used by pilot lights increases the amount of energy used by a gas range. Given a range with three pilot lights (one for each pair of burners on the top of the unit and one for the oven), cooking for a family of four for one month will use about 820,000 Btu, just under twice the energy used by an electric range. However, in the winter the heat produced by the pilot light is released into the room and cannot be considered a waste of energy. Similarly, in the summer the heat produced by the pilot lights must be removed by the cooling system, increasing the energy waste. Hopefully, American manufacturers will soon produce gas appliances with automatic electric igniters instead of pilot lights. European manufacturers have been doing this for many years.

Adjust the flame on gas cooking units so that it burns a steady blue, with only an occasional fleck of orange. This adjustment, best made by a serviceman, will result in the most efficient combustion.

Adjust the pilot lights to the smallest flame that will reliably light the burners and that will not be blown out by normal room drafts.

Use the correct pan and burner combination so that the pan completely covers the ring of flame. A pan that is too large may cause excess heat to flow along the stove top and discolor or crack the enamel finish.

Since most of the heat transfer from a flame is by radiation, dark-colored pans are best.

Self-Cleaning Ovens

There are two types of self-cleaning ovens: the "continuous clean," or catalytic, oven and the "clean-cycle," or pyrolitic, oven. The catalytic oven uses a special coating on the oven liner, which breaks down oven residues at normal baking temperatures. It does so any time the oven is used, but the oven is never completely clean at any one time because it takes several hours to decompose any one splatter, and by that time others have usually accumulated. Because all cleaning is done during normal use, no additional energy is used for the cleaning process.

The pyrolitic oven uses a high temperature, about 800 to 900 degrees, for a period of about 2 to 3 hours to burn any food residue off the oven parts. A fine ash is all that is left at the end of the cleaning cycle, and the oven is completely clean. In an electric oven the high heat is produced by conventional baking and broiling units in combination with a special heating element around the outside shell. The smoke produced by the burning-off process is removed by a small electronic air cleaner installed in the oven vent. A circulating fan blows air around the oven shell to keep the range surfaces from getting too hot to touch, and an automatic lock prevents the oven door from being opened at dangerously high temperatures. In a gas-fired pyrolitic oven the process is much the same, with the gas flame producing the heat. These ovens do use a considerable amount of energy (about 50,000 Btu for gas ovens and about 12,000 Btu for electric models) during the cleaning cycle. For this reason, clean pyrolitic ovens as seldom as possible, and do it at a time of day when the heat will be most beneficial during the winter and when it will do the least harm in the summer (usually the early hours of the morning in both cases). Most self-cleaning ovens have a timer that can start and stop the cleaning cycle automatically.

Microwave or Electronic Cooking

An electronic cooking unit, whether built into a conventional appearing range or installed as a separate counter-top appliance, cooks food by using high-frequency radio waves—about the same as radar—to violently shake water molecules, which generates heat. Of course, these ovens will only heat items that contain moisture; they will not heat dishes, pans, and so on. The presence of metal shields the food from the radio waves, so foil and metal pans should not be used.

The efficiency of the microwave cooking system in converting electrical energy into microwave energy, which is then absorbed by the food, is about 45 percent. However, because only the food is heated, and water or air is not needed to transmit the heat to the food, it may be much more efficient than surface or oven cooking. For example, if the object is to heat two cups of water, the range unit will be more efficient. If, however, the object is to cook two hot dogs, it is much more efficient to do it in a microwave unit because there is no pan or water to be heated. Both a medium-size microwave unit and the large burner on an electric range use about the same amount of power—1,700 watts. To cook two hot dogs in two cups of water on the surface unit would take about 5 minutes; two hot dogs in buns in the microwave would take about 40 seconds. Similarly, a beef roast that would require 45 minutes in a conventional oven, using 2,000 to 3,000 watts for 10 minutes to preheat and then about 1,700 watts applied intermittently to maintain the temperature, would cook completely in a 1,700-watt microwave unit in 12 minutes.

Microwave units with browning elements waste considerably more heat, and cooking food in a microwave oven and then browning it under a 5,000-watt range broiler is not conserving energy either. Because an electronic oven does not get hot, spatters are not baked on and high temperature oven cleaning is not necessary. Microwave ovens do not replace conventional ovens for many special uses, however, particularly the baking of pastries. Microwave cooking of foods not requiring water may be up to four times as efficient as range-top cooking in terms of the total energy used.

Cooking Management

Plan to cook several items that can be baked at the same temperature for the same meal or for future use; it takes only a small amount of additional energy to cook several items rather than to cook just one.

When the oven is hot, use it—preheating uses a great deal of energy.

As mentioned earlier, turn the oven off about 10 minutes before the food is done—the residual heat will finish the cooking.

Plan combination dishes that can be cooked in one pan on one burner.

Single-Purpose Cooking Appliances

The various single-purpose cooking appliances common to a household, such as toasters, coffee makers, waffle irons, and so on, are usually just as efficient if not more so than the conventional range, because they are specifically designed for one purpose and can incorporate automatic controls and special features that cannot be built into the general purpose range or oven. The one possible exception to this statement is the tabletop oven or broiler (not microwave), which may be more efficient for very small items or short cooking times, but is not as well insulated as the range oven and will lose more heat to the room when used for long periods of time.

Dishwashers

The dishwasher consumes energy in three ways: a motor circulates the water during the wash cycle, a heating element is used to raise the water temperature and to dry the dishes, and the hot water used must be generated by another system in the house. A dishwasher does not use any more hot water than hand washing might if it is used only when it has a full load. It will usually hold a full day's supply of dishes for a family of four and will use only about five gallons of hot water to clean this load. A housewife doing dishes after each meal will use at least that much if not more. The higher temperature and stronger detergent used in dishwashers will also get the dishes cleaner.

Run the dishwasher only when there is a full load. Open the dishwasher and let the dishes air-dry rather than using the drying cycle. This can cut the normal energy consumption in half. The dishes should be hot enough to dry quickly in any case.

REFRIGERATION

The household refrigerator and freezer have made major contributions to the high standard of eating enjoyed by the current generation of Americans, but they have also become major consumers of energy, much of it in the name of convenience.

Refrigerators and freezers work on exactly the same principle as do air conditioners. Because the refrigerator and freezer operate at much cooler temperatures than the air conditioner, when the air in the room hits the cooling surfaces, moisture condenses and frost forms. It is our attempt to remove this frost that complicates the energy picture.

Refrigerators and freezers are available in three major types—manual defrost, automatic defrost, and frost-free. The manual defrost is just that—frost accumulation on the cooling surfaces must be removed by manually shutting off the cooling unit and allowing the frost to thaw, using either room air, heat from pans of hot water, or special defrosting heaters.

The automatic defrost unit, usually available in refrigerators only, is actuated by a timer, and the refrigeration unit shuts off for a period long enough for the frost to melt, and the resulting water drains into a collecting pan near the condenser coils. The heat of the condenser then reevaporates this water into the room.

The frost-free refrigerator or freezer uses concealed cooling plates and an automatic fan that blows air across these plates and through the refrigerated area. On a timed cycle, electric heating elements melt the frost from the cooling plates and reevaporate it. Electric heating elements are needed because the freezer cannot be allowed to become warm enough to melt the frost.

Both automatic defrost and frost-free refrigerators often have heating elements around the doors to melt any frost that may form there from air leaks in the door gasket. This, along with butter keepers and other specially warmed sections of the refrigerator, adds to the amount of heat the refrigeration mechanism must remove.

Finally, bowing to the great American need for styling, the refrigerator manufacturers have hidden the condenser coils under the machine rather than mounting them in an exposed position on the back as was done earlier. This means yet another motor and fan in order to blow air past this concealed coil, rather than allowing room air to flow over the coils by convection.

As a matter of comparison, in a 14 cubic foot refrigerator the cooling unit of a manual defrost refrigerator uses about 350 to 400 watts and operates 40 percent of the time. In a frost-free model the cooling unit uses 615 watts and operates 40 to 60 percent of the time.

A manual defrost or automatic defrost refrigerator uses about 40 percent less energy than a frost-free model.

If your refrigerator has a switch for the door heating elements (usually marked humid-dry), turn it to the "dry" position, which will shut off the heaters, unless you see that frost actually starts to form around the door. This can save up to 16 percent of the operating costs of the unit.

Turn off the "extra" refrigerator in the basement, if you have one, unless it is really needed; you can save $3 to $5 per month, depending upon its type and size.

Open the refrigerator or freezer door as little as possible. Collect the foods you wish to put in or take out on the counter near the door,

and then open the door once. The greatest heat loss occurs in the first few seconds that the door is open.

Make up your mind what you need before you open the door—contemplation may be the joy of philosophers, but not in front of an open refrigerator.

On manual defrost models, defrost frequently. Frost is not a good heat conductor, and frost buildup will reduce the unit's efficiency.

Clean the cooling coils periodically, whether they are behind or beneath the refrigerator or freezer. Heat must be transferred to the air, and a layer of dust acts as an insulator. Use a vacuum, the blower end of a tank-type vacuum cleaner, or a suitable brush.

OTHER APPLIANCES

Televisions

Color television sets use about twice as much energy as black and white televisions of the same picture size. Solid-state or transistorized sets use ⅓ less than tube-type units in color versions and ⅔ less in black and white. Fancy features such as wireless remote control use even more, not because of the power it takes to change the station but because the receiver for the channel changer must be on continuously. "Instant-on" sets, which produce a picture immediately, are able to do so because most of the set is *always* on. They are still using a great deal of energy while turned "off," and this can double or triple the annual power usage.

When you are not watching the television, turn it off.

Use a black and white set when color really does not make a difference.

If you have an "instant-on" set, put a switch in the power cord so that it can truly be turned off.

If the set is portable, use it on the porch in the summer so that the heat gain does not add to the cooling load.

Read a book instead.

Other Small Appliances

Other small appliances, such as an electric can opener, carving knife, toothbrush, pencil sharpener, and so on, really use so little power that their efficiency, use, or nonuse becomes immaterial. It is doubtful that if the power used by all these small appliances in the average home was totaled for a year it would exceed 100 kilowatt-hours or about $3 worth of energy. Other savings are far more significant.

Quality of Product

The production of any appliance involves energy, and in great amounts when one considers the fuel requirements for making steel sheets, wires, and so on. If appliances can be made to last longer, to require less servicing or replacement of parts, then considerable energy will be saved. This means that greater emphasis on product quality, even at somewhat greater cost, is a step in the right direction. Certainly, the concepts of planned obsolescence and disposable products are not geared to the problems of a future society of scarcity.

12

Play It Safe

THE AIR SUPPLY

The most important safety factor in burning fossil fuels (oil, gas, coal) is an adequate air supply to provide oxygen for the complete burning of the fuel. If there is enough oxygen present, the carbon contained in the fuel is converted into carbon dioxide, which is a harmless gas. If there is insufficient oxygen, carbon monoxide is formed. This is an odorless, tasteless, and deadly gas. A central heating system should have a fresh air supply, preferably taken from some area outside the living quarters. This fresh air supply may be an open vent in an unheated crawl space, a vent pipe extending from the furnace compartment to a well-ventilated attic, or from a basement or living area with good air infiltration. An extremely tight house should always have a separate fresh air supply to the furnace room.

HOUSEKEEPING

Probably the most frequently violated safety rule is: *Do not use the furnace room for storage.* Any flammable material stored there becomes a potential fire hazard if the heating system malfunctions in any way. It is best to line the furnace room walls with cement—asbestos board rather than ordinary dry wall.

FUELS

Oil

There are two potential hazards connected with the use of oil as a fuel. The first is the matter of leaks. If oil leaks out on the basement floor, some of it may vaporize and be ignited by a pilot light on a gas-fired appliance, or some other source of heat. This is a greater hazard with No. 1 oil than with No. 2, since No. 1 oil vaporizes much more rapidly. The second hazard is that of improper ignition in the oil burner. If the flame is slow to ignite, too much oil will be in the burner at the time that ignition occurs, and it will ignite with a puff or bang. The pressure from this small explosion may loosen the joints of the flue pipe, blow the burner door open, or otherwise damage the heating unit or even open it to the point where some flame might escape and cause a fire in adjacent materials. If your oil burner starts with a bang, call a serviceman. Do not attempt to fix it yourself.

Gas

Both natural and liquefied petroleum gases are odorless as they are drawn from the ground or manufactured from oil. However, a substance with an odor is added to help detect the presence of leaks. It takes only a small concentration of gas in the air to create an explosive mixture—and an explosion can even demolish an entire house. If you smell a strong gas odor, leave the house and call the gas utility company or your local fire department. Both have equipment for determining if the gas mixture in your house is potentially explosive, and can ventilate the area. Do not turn any light switches on or off. The potential hazard of natural gas is less than that of propane because natural gas is lighter than air and tends to rise toward the ceiling and dissipate in the room. Propane gas is twice as heavy as air, and tends to flow along the floor and gather in low places, where a spark can touch it off with disastrous results.

If a slight odor of gas is detected, the best way of locating the leak is to brush any suspected joints in the gas piping with the bubble-producing liquid used by children. Any small leak will immediately create bubbles and make itself obvious. A soap solution can be used, but the bubble-producing liquid is better.

All gas pipes and connections should be handled by an experienced plumber who has both the knowledge and the special equipment to handle gas safely.

Liquefied petroleum gas should never be stored inside the house in any quantity larger than the small 1 pound cylinders used to operate torches.

All gas appliances should bear a label indicating that they have been approved by the laboratories of the American Gas Association.

Electricity

All electrical installations should be made according to the provisions of the National Electric Code. If you are not familiar with the provisions of this code and do not have experience in working with electrical wiring, don't be a do-it-yourselfer. Call an electrician. Electricians may be expensive, but so are funerals.

If your home is heated with electric resistance baseboards, be sure that you do not restrict the airflow over the baseboard with draperies or curtains. Though a hot-water convector or forced-air heat outlet can only get as warm as the water or air circulating through it, the electrical convector will continue to get hotter if the airflow through it is restricted. It may reach a temperature that could ignite the curtains or the wall.

Know the location of your fuse box or circuit-breaker panel, and be sure that all fuses or breakers are labeled so that you can quickly turn off the power to any appliance or circuit that shows signs of being defective.

Never replace fuses with ones that are larger than the circuit is designed for. In ordinary residential wiring the only circuits that should have larger than 20 ampere fuses are those that serve major built-in appliances.

If you have any doubt about the safety of your electrical system, call your city electrical inspector, your local power company, or a licensed electrician.

Be sure that a portable electric heater has a built-in switch that turns the heater off if it is tipped over. An electric heater is certainly hot enough to start a fire if tipped over by a child or a pet tripping on the cord or stumbling over the appliance. All heating appliances and the cords to them should be placed out of reach of children and pets.

FIREPLACES

Fireplaces have three inherent hazards. The first is that of a chimney fire, particularly if coal is burned. The chimney should be checked for accumulated soot. Though a chimney fire can be a frightening experience, it is usually not dangerous to the structure unless the mortar joints in the chimney have deteriorated.

The second danger of fireplaces is damage from sparks or flames coming from the front of the fireplace out into the room. They may be caused by wood that pops or by overfiring. The latter is especially common when trying to burn packing materials, paper, and old Christmas trees. Some of these materials burn almost explosively and produce so much flame that it cannot all go up the chimney, and some comes out into the room. *Never* use kerosene, gasoline, or charcoal lighter fluid to start a fire in a fireplace. Use care in firing such materials in small quantities, and equip the fireplace with an appropriate fire screen to prevent sparks from popping into the room.

The least known danger of fireplaces is the fact that the backside of the fireplace masonry can become very hot when a fireplace has been burned hard for a long period of time. This can be a real problem when the fireplace is used as a primary source of heat due to a power failure. The back of the fireplace can easily become hot enough to blister paint and possibly start a fire in combustible materials that are stored against the back wall of the fireplace in another room. If your fireplace is to be fired hard for long periods, be sure that all the masonry is clear of combustible materials.

EMERGENCY HEATING

In case of power failures or fuel shortages, extreme care should be taken in preparing emergency heat. Gas ovens and burners should not be used for long periods to heat the house. Although the combustion products given off by an oven or burner are fairly insignificant in the quantities normally found in cooking, an attempt to heat the house with unvented equipment will discharge entirely too great a quantity of waste gases into the house and may result in danger to the occupants. Gas or gasoline-burning appliances can be used for short periods of time in relatively well-ventilated houses.

Never use charcoal in an attempt to heat the house. Charcoal invariably gives off large quantities of carbon monoxide when it burns, and dozens of fatalities can be attributed to its misuse as a source of emergency heat. It is possible to burn charcoal in a fireplace with a functioning chimney, but nowhere else. Even a small hibachi can produce enough carbon monoxide to be fatal.

Catalytic heaters of the type used by campers are safe to use in emergency situations. Other gasoline-burning appliances used by campers are intended for use in the great outdoors or in tents or campers that have considerable air infiltration. They are not intended for use in a tightly closed house and should not be used there.

ELECTRIC BLANKETS

Electric blankets should not be used on children's beds. Shorts can develop in electric blankets, and hot spots can be formed if the blanket is crumpled up and turned on. Several heated layers, when pressed together, can generate enough heat to start a fire. Electric blankets should be handled with care, laundered according to instructions, and spread over the bed without doubling. They should be turned off when not in use. A lightweight blanket over an electric blanket increases its efficiency.

MIRACLE WORKERS

The energy crisis is certain to stimulate the ingenuity of the American inventor. Though some useful devices may well be produced, there will undoubtedly be a number of gadgets marketed either in the spirit of fraud or ignorance that will be at best useless and at worst dangerous to the buyer. Just as in the case of all the accessories for an automobile that increase your gas mileage by 10, 20, or 50 percent, new gadgets that are supposed to save you 10, 20, and 50 percent on your fuel bill should be equally suspect.

One example of a new device that is being promoted as an energy saver is an automatic damper that opens the chimney flue when the furnace goes on and closes it when the furnace goes off. This device will indeed prevent warm indoor air from going out the chimney when the furnace is not in operation. There is only one problem—the device is not fail-safe. Flue gases are hot and to a certain extent corrosive. If this device should malfunction when it is in the closed position, all the combustion products from the furnace would go into the house, quite possibly with fatal results.

No device should be attached to your heating system unless it is approved by the manufacturer of that heating system or by the laboratories of the American Gas Association. This organization was set up by the gas distributing utilities and the manufacturers of gas appliances, and has contributed greatly to the enviable safety record of that industry.

A second type of gadget that may soon make its appearance is a device to reduce the fuel input to your heating system, be it gas or oil. Reducing the fuel input below manufacturers' specifications is likely to increase ignition problems. Also, tests have shown that the mixing of gas and air in a gas burner is so precisely calibrated that significantly reducing the gas input will actually reduce the carbon dioxide content and the combustion efficiency.

Beware of the energy miracle worker—in addition to your money, your life may be at stake.

13

Home Design Considerations

Most of the items discussed up to this point are applicable to existing houses and apartments. Consider now the future. With energy conservation a significant problem in the forseeable future, how will this affect the construction of housing? Not only the house that you would like to build, but housing in general. Should we keep on moving farther and farther into the suburbs? Or should we try to reverse the trend and revitalize the inner cities? Have other older nations given us some indications of what might happen in this country?

SINGLE-FAMILY AND MULTIFAMILY DWELLINGS

The American ideal is a single-family dwelling with an acre of land around the house. With the rapid increase in population, and with the work force moving from agricultural areas to the cities, the trend has been toward more people living in a smaller number of metropolitan areas. For a variety of reasons this has meant that more and more agricultural land was being claimed by the suburbs to house the working population. The practice of commuting long distances to work has not discouraged people from moving still farther away from the central city.

With gasoline shortages likely to remain with us in the future, there may have to be some rethinking as far as housing trends are concerned. We could move toward individual housing on smaller and smaller lots, with their fences and shrubbery at the boundary lines to preserve an "island of privacy," as many foreign countries have done, or we could move toward multihousing units, whether they be apartments, duplex residences, or row housing. The renewal of inner city housing will occur if public transportation to the suburbs is lacking, fuel for individual commuting is scarce, the cost of individual housing becomes too great, land prices increase too rapidly, or traffic congestion reaches an intolerable level.

We might also move toward the concept of dispersed cities, attempting to reverse the trend toward the formation of a regional megalopolis, returning to the United States of earlier years. It will require less heating and cooling energy to live in multifamily housing units. An apartment, sandwiched between others above, below, and on both sides, will have an outside exposure on only one of its six sides. The top and end apartments might have an additional exposure or two, but none will have the full exposure of an individual dwelling.

TWO-STORY VERSUS ONE-STORY

The thermal efficiency of a one-story house is not necessarily any greater than that of a two-story house.

For example, consider two houses with the same 2,400 square feet of floor area—a single-story house 40 by 60 feet and a two-story house 30 by 40 feet. Assume both have a 30 by 40 foot basement. The gross wall and ceiling areas are as follows:

Gross Areas	*One-Story House (square feet)*	*Two-Story House (square feet)*
Exposed wall	1,600	2,240
Exposed ceiling	2,400	1,200

The one-story house has more ceiling area, but this is the easiest area to insulate. Depending upon the degree of insulation that is applied to walls and ceiling, one can prove anything one sets out to prove. Obviously, if the one-story building was not properly insulated, it would prove to be a thermal disaster. However, if both houses were insulated to R=14 in the walls and R=22 in the ceiling, the heat loss would be about the same.

SQUARE AND RECTANGULAR HOUSING

For a given floor area a square floor plan has less wall exposure per square foot of room area than a rectangular floor plan. A long structure like a mobile home has a relatively large wall exposure.

SMALL AND LARGE FLOOR AREAS

A large building has, proportionately, less wall exposure per square foot of floor area than a small building. A large building is, therefore, easier to heat efficiently. Of course, the smaller building has a smaller total heat loss, and is less wasteful of energy in absolute terms. The true measure of efficiency in this case may be in terms of the number of the building's occupants, the number of units in the building, and so on.

WINDOW TO WALL AREA

In general, the more windows the greater is the heat loss. However, it does make a difference where the windows are located, especially in terms of solar heat gain during both summer and winter.

Building codes usually specify a minimum window area of 10 percent of the floor area. For example, for a floor area of 1,000 square feet, at least 100 square feet of windows are required, corresponding to about seven average-sized windows.

In general, east and west windows are efficient sources of heat during the summer, while north windows receive the brunt of winter winds. This leaves the south side as the preferred location for windows from the standpoint of solar heat gain.

THE ORIENTATION OF THE BUILDING

A building whose long dimension runs north and south will receive more solar heat in the summer than the same building whose length runs east and west. With two identical buildings, 20 by 50 feet, the amount of heat received by the walls from the sun was found to be 80 percent greater for the building with the north-south axis than for the one with the east-west axis.

SOLAR ORIENTATION

Because of the position of the sun in relation to our part of the earth, the south wall receives a relatively small amount of solar energy in the summer compared with that received by the same area facing the east or west. For example, on a summer day the maximum solar energy received on the east wall at 8:00 A.M. and the west wall at 4:00 P.M. is about 220 Btuh for each square foot of surface. The maximum received by the south wall at noon is only about 90 Btuh per square foot. This is less than half, even when the total heat gain for the day is taken into consideration.

In the winter these conditions are reversed. The heat gained by the south wall in the winter is much greater than the same wall in the summer, because the solar radiation hits the wall at almost right angles.

The fundamental importance of the south wall in the solar heating of dwellings was first recognized by George Fred Keck, who promoted the concept of solar orientation. The windows on the south wall should be protected by a roof overhang designed to prevent the sun from striking the window in the summer, but which will permit the winter sun to shine into the room and warm it.

The roof overhang as a sun control does not work with east or west walls, so that other forms of shading are required.

Since the summer sun striking the east or west walls can impose a large heat gain, one possible means of reducing the cooling load would be to locate the attached garage at the east or west side of the house.

For the same reason a wide porch on the east or west side of the house would serve as an effective permanent awning. The reduction in solar heating from the winter sun on these walls is negligible.

A wide porch on the south wall would not be advisable because it would shut off the winter sun, whereas a special roof overhang would not.

CONSTRUCTION BELOW GRADE

A few feet below the surface the ground temperature is relatively constant and attains the average annual temperature of the region. The deep soil temperature will range from 50 degrees in the north to about 60 degrees in the south. Any wall exposed to the ground, therefore, will have a constant temperature mass on one side. Regardless of the wall construction, the heat flow will be extremely small, since the temperature difference between the basement air and the ground is small. Furthermore, the heat transfer between the basement air and the ground does not change much from winter to summer. Therefore, it is not necessary to insulate a basement wall more than 2 or 3 feet below the surface.

As an extreme example, a house built entirely below the surface of the ground would probably not require much of a heating system because internal sources of heat would probably satisfy the needs of the occupants, especially if the roof were insulated or covered with earth. In fact, it is probable that cooling would be required even in the spring and fall.

The split-level house is frequently built with the lowest story partly enclosed by earth protection. A hilly terrain creates some interesting architectural possibilities in terms of building rooms against the ground and increasing thermal efficiency.

TRIPLE GLASS

In examining all the sources of heat loss in a house, the item that appears to offer the greatest possibility for improvement is the window. The practice of using double-paned glass will undoubtedly spread to the warmer areas of the country. At the same time the use of triple-paned glass appears to be practical in areas where double glass is now common.

WIDER STUDDING

The possibility of using 2- by 6-inch studding in place of the common 2 by 4 needs investigation. This would leave more space for wall insulation.

WINDBREAKS FOR WINTER

The role of trees and shrubs in energy conservation cannot be overemphasized. The farmer isolated on the prairie has long recognized the importance of windbreaks. In a city the neighboring structures serve as windbreaks, but a study of the individual house and lot may reveal where effective plantings could be made.

PUT IT ALL TOGETHER . . .

If all the foregoing ideas were incorporated into one building, it might be filled with contradictions. It would be square. It would be small, with no more than a 10 percent glass-to-floor-area ratio. The long axis (if there was one) would run east-west, it would be partly below ground, the garage would be on the east, a wide porch would be on the west, a roof overhang would grace the south. It would have triple glass windows, perhaps 2- by 6-inch studs in the walls and it would probably be so ugly that no one would want to live in it.

PART TWO
THE AUTOMOBILE

14

Driving for Economy

Conserving the nation's petroleum resources by judicious use of the automobile is not exactly a new idea. According to an advertisement in *Collier's Weekly,* December 16, 1916, there was concern even in those days about "cutting down the nation's fuel bill." The solution was to use Goodyear Cord Tires (today it is radials). The advertisement also contained a prediction from the scientists of the day that the nation would exhaust its gasoline supply in twenty-seven years, and "were every one of the three million American cars equipped with Goodyear Cord Tires, this supply would last seven years longer, or thirty-four years."

Such a forecast is amusing in view of what we know today, yet there's no denying that the world's supply of fossil fuels is finite, which means someday we will, indeed, run out of gasoline. Long before that day arrives we hope that some alternative energy source will be developed and proven practical as a means of propelling the world's automobiles.

In the meantime all of us must come face to face with a worldwide energy crunch. Its direct effect on the motorist is obvious. We are being asked to make do with less. And that is where these pages should be of value to every driver. We have collected fuel-saving ideas from a variety of sources, including professional economy drivers who were the winners of the famous Mobil Economy Run.

There are limits to an automobile's capabilities, and when approaching those limits it may become exceedingly tedious for a driver to strive for the ultimate economy. On a practical level, however, the fact is that most drivers can with relatively little effort improve fuel economy by amounts of 30 to 40 percent. (A nervous driver, truly careless in his driving habits, can do even better once he becomes retrained in economy techniques.)

An automobile's mechanical condition has a great deal to do with economy, and we will discuss these factors so that even the non-mechanical driver can comprehend them.

We cannot guarantee to make you an overnight expert in the daily economy run, but the advice that follows has been tested and is valid. Some items may seem minor, but the important thing is to remember that saving energy while operating an automobile is a combination of many factors that range from driver technique to the condition of the car.

HIGHWAY CRUISING TECHNIQUES

Up to a point, slower highway speeds mean better fuel economy in the average car. For the time being the entire United States should be driving at no more than 55 mph on the open road. And it is on the open highway where the 55 mph limit will save the motorist considerable fuel over the former legal speeds; in some cases they were 80 mph. Boring as 55 mph driving is on an interstate highway, economy from a typical V-8 engine will increase dramatically as speed is decreased.

Of course, no two cars or drivers are exactly alike, and there are exceptions to this slower-is-more-economical rule. Occasionally a full-sized vehicle, equipped with an unusual drive train, can return phenomenal fuel figures in the 60 mph range. There are some sports cars with aerodynamic bodies and five-speed transmissions that actually get better economy at speeds above 60 mph. However, the vast majority of cars today are boxlike in shape, equipped with an automatic transmission, a good-sized V-8 engine, and some type of exhaust emission control system. This average car normally returns the best miles per gallon at a steady driving speed of about 45 mph. Below 40 the engine may be "lugging" in top gear, and the economy will drop off to a surprising degree. Above 40 and up to 45 optimum economy conditions will be found, while the number of miles per gallon will begin to drop around 50. Each 10 mph increase in speed will show a graphic decrease in miles per gallon of gasoline. At 60 or above the carburetor is generally working full time instead of part time. This fact coupled with the greater aerodynamic drag as the car pushes its way through the air accounts for most of the increase in the use of fuel at higher speeds.

EFFECT OF TYPE OF DRIVING ON FUEL ECONOMY

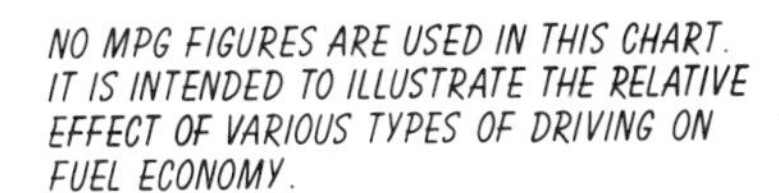

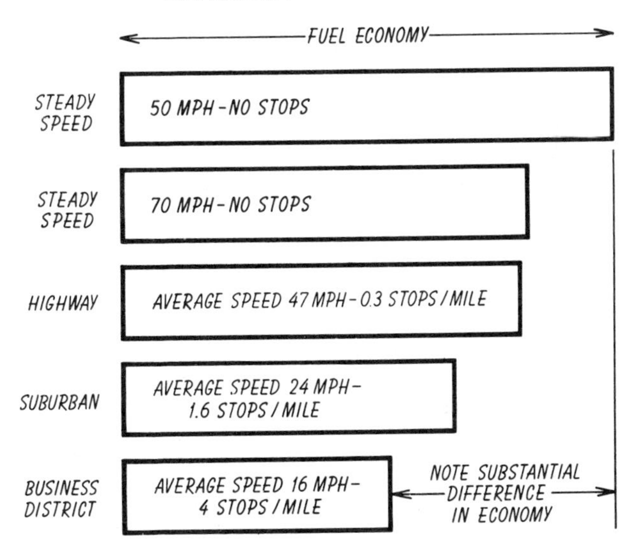

Here is an example of constant speed fuel economy taken from a test of a 1974 Chevrolet Vega subcompact. Because it was equipped with its basic engine and an "economy" rear axle ratio, the figures support the slow-is-economical rule. We offer them to demonstrate how rapidly gasoline flows as speeds increase. At a steady 70 mph, gas consumption is 22.33 mpg; at 50, 30.52; and at 30 mph an incredible 42.76 miles per gallon. These are steady speeds; there is no allowance for starting and stopping.

Until recently most people drove about 70 mph on the Interstate Highway System, which was designed for just such high-speed travel. Late model cars at these speeds return from 10 to 15 miles per gallon, not very efficient in terms of fuel, but great in terms of travel time on the road.

In 1973 the summer fuel crisis prompted the respected auto magazine *Road and Track* to conduct an economy test under controlled conditions. A new Chevrolet Camaro with a 350-cubic-inch V-8 engine developing 245 hp (the high performance version) and a four-speed manual transmission was used. Driving a test route of highways, urban and suburban roads, the Camaro averaged 11.2 mpg. All tests were conducted without using the air-conditioning system. The Camaro was equipped with a flow meter to compare constant speed driving beginning at 30 mph and moving up in 10 mph increments to 70 mph. *Road and Track* found that the Camaro used more gas in each step after 40 mph, which was expected. Comparing 50 and 70 mph, the Camaro used approximately 25 percent less fuel at 50.

These figures are encouraging, but there is a penalty—the time lag. It takes 34.3 minutes longer to traverse 100 miles at 50 mph

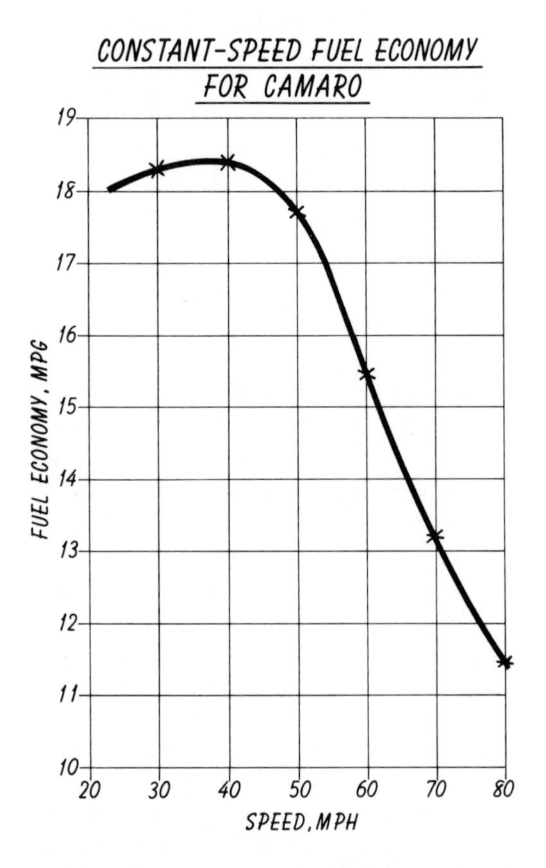

than it does at 70 mph. The Camaro averaged 17.7 mpg at 50 and 13.3 at 70. It is also interesting to note that the high performance car did better by more than 2 miles per gallon at a steady 70 on the open road than it did in stop-and-go driving.

Today the 70 mph cruise speed is no longer legal, and most drivers will be pleasantly surprised at the miles per gallon returned by their cars on their next highway trip. In addition to reduced speeds there are other techniques that can be employed to increase the good economy figures even more. A gentle pressure and steady foot on the gas pedal will do wonders—a couple of miles per gallon in fact—on the open road as opposed to the pedal-pumping style of driving that is most common.

Ideally, try to find a throttle setting that is comfortable to maintain and that also puts the car at 55 mph on a level road. The foot must be comfortable in order to maintain the same position for miles on end. Once you find the correct position, try backing off the gas very gently to see if the pedal pressure can be reduced slightly without losing any road speed. It sounds crazy but on full-sized cars in particular it seldom takes as much throttle pressure as was needed to get up to speed to hold that speed on level ground. Backing off the pedal just a shade can save gasoline without costing a minute in time.

Keep a respectable distance from the car ahead on the highway, ensuring enough room to keep a steady foot on the gas without having to change speed or use the brakes to accommodate other traffic. Professional economy-run drivers aim for a throttle setting and speed that requires no change at all. This becomes slightly paralytic after a few hours, but it does produce great fuel economy figures on standard automobiles. The idea is that each time the throttle is moved, the accelerator pump is activated and squirts a tiny bit of gas into the carburetor.

The normal driver will not want to maintain the rigid discipline of economy-run competition driving on highway trips. However, just maintaining a relatively steady foot pressure and about 55 mph on a long haul will produce up to 5 miles more per gallon than driving the same car at 70 on the same road.

Use gentle acceleration techniques on the open road, and avoid ramming the pedal to the floor after a traffic slowdown. Wide open acceleration is wasteful in terms of gas in any situation with almost any type of car or equipment. If the highway goes through rolling country, gasoline may be saved by holding a constant pedal pressure on the inclines. In other words, allow the speed to fall off a little as the car climbs, holding the throttle steady. The speed will increase on the down side, and very little time will be lost using this technique. Each automobile drive train and gear ratio has an ideal climbing speed, but in general the constant throttle is the most economical way to drive in rolling country.

A good tip for highway fuel conservation is to avoid playing the gas pedal to keep right on the speed limit. Let the car seek a comfortable cruising speed in the neighborhood of the speed limit. Incidentally, be sure you have an accurate speedometer, or at least know how to compensate for the degree of error in the instrument. Travel time will increase while running at 55, and most of us will want to drive right at the limit for a long trip.

You can check the accuracy of the speedometer on the highway and find a fairly accurate reading for 55 mph. Use your odometer or the measured mileposts that many states post on major highways. The equation to find elapsed time for a given speed is simple. Divide the speed into 60. and the result is the minutes per mile factor in decimal form. For ease of checking against a watch or the car's clock, translate the hundredths of a minute into seconds. Here is how it is done: It should take exactly 1 minute to go 1 mile at 60 mph. For a 50 mph speed, the factor is 1.2 or 1 minute 12 seconds for each mile covered. At 55 there is a factor of 1.091 or 1 minute and a fraction over 5 seconds for each mile covered. At 55 mph it should take exactly 12 minutes to go 11 miles. Keeping a steady foot on the throttle, and knowing how fast you really are traveling, can make your highway journey an economical one and as swift as legally possible.

Cruising along an interstate highway usually gives the driver plenty of room to practice fuel conservation techniques. Less modern four-lane highways and two-lane byways bring on more problems with traffic. It is often difficult to maintain a decent space between cars on congested highways, for when the courteous driver allows a gap to form, invariably someone will pass and jump into the slot.

Traveling at high speeds in close company does provide a fuel economy bonus, but it is one form of fuel saving that is both illegal and dangerous. When tailgating at speeds above 50 mph you may catch a "draft," especially if you are following a big truck. If the car has a tachometer, you can see the rpm drop as you get close to the truck while your speed remains the same while trailing in his wake.

Drafting is a technique used by race drivers to build up enough speed to pass, and drafting a truck on the highway will increase fuel economy. The hazards should be clear; if you are close enough to the vehicle ahead of you to benefit from the draft, you are too close to suit the vehicle code. Drafting is not recommended for the private citizen on the open road.

The main rule for achieving fuel economy on the highway is to be steady and gentle on the gas pedal. The steady foot and the maintenance of a constant speed will return excellent fuel figures at almost any road speed. Consistency is the secret. Poor results in the 9 to 11 mpg range on big cars are only partly due to high average speeds. To maintain those high speeds the average driver does a lot of slowing down and speeding up, often accelerating heavily uphill, a real gas-eating habit. Running along at a constant rate of travel, be it 50, 60, or 70, gliding up inclines and coasting down the other side, these are the real tricks and the basic methods for picking up extra miles per gallon on the highway.

THE TAKEOFF

The simple act of moving a car from rest can be either thrifty or extravagant. The normal method of powering away from a stop sign is somewhere between the creep and the drag race start; the latter is plain awful for fuel economy. Full throttle acceleration is close to a complete waste of gasoline. The car does get up to speed rapidly but a large part of the fuel gulped through the carburetor with a foot to the floor is not totally burned in the combustion process; it is tossed overboard and out the tail pipe. The banzai dash is a thing of the past for the economy-minded driver.

On the other hand, a closed throttle creep in the modern car is not the real answer to saving gas either. The most economical method of getting under way is to get the car in top gear as soon as possible.

Driving an automatic transmission, almost standard today with the majority of cars, it is not difficult to learn to hear and feel the shifts under gentle acceleration. Practice will tell you just how much throttle is needed to get into drive range, usually somewhere between 20 and 30 mph. Use a gentle, light pressure on the gas pedal to get moving initially, then gradually increase the throttle pressure until you feel the gears shift into second and finally into drive. The car's motion will probably be slower than usual, and it will take another hundred feet or so to reach a speed of 35 mph. However, you will have saved many ounces of gasoline. Powering away from a stop not only wastes gasoline in poor combustion, but the automatic transmission will remain in a lower gear longer, causing longer running with higher rpms on the engine.

Economical driving with a stick shift also involves the idea of getting into top gear as soon as possible. Take a tip from the taxi drivers who favor the basic three-speed manual transmission for economy of operation. In city traffic most cab drivers will start in second gear and drop into high very quickly, perhaps lugging the engine a little, but not so much as to be harmful over the long haul. If the three-speed manual transmission is coupled to a low horsepower engine, it might work better to start in first, run up a good rpm, and drop into high, bypassing second. Should the car be struggling up an incline use all three gears but pass into high gear as quickly as possible.

The smaller engined compacts and imported cars often are sold with a standard four-speed manual transmission. Depending on available engine torque, the car may be started in second gear and dropped directly into fourth. On less powerful engines, start in first, go right to second, build up a little road speed, bypass third, and go from second to fourth. The technique of shifting the four-speed transmission for economy depends a good deal on what feels most comfortable for that particular car and engine. Violent bucking caused by lugging the engine is not only hard on the clutch and engine, but it does not save any fuel. The ideal situation is to keep the engine running smoothly while keeping the rpm as low as possible through the gears.

It must be emphasized that the small four-cylinder engines in many cars will not tolerate continual lugging. These engines return much better than average fuel economy anyhow, so be sure to provide enough rpm at all road speeds to keep the engine functioning properly. Merely avoiding the full throttle type of start with this size automobile will save a considerable amount of gasoline.

Avoid the jackrabbit start at all costs. Gentle and steady throttle pressure will get the job done on any vehicle and produce better fuel economy.

HOW TO STOP

No matter how well the driver figures his route and traffic there comes a time when he must stop. It is important to know methods of stopping that will conserve fuel. Primarily the idea is to use the brakes as little as possible in order to conserve energy. Stepping on the brakes generates heat in the linings and heat is energy. To manufacture that energy the automobile has burned gasoline. Dissipating heat/energy through the use of the brakes uses gasoline because fuel is burned to replace that energy in order to propel the car. Of course, the brakes must be used and some energy expended to stop, but in many cases a good deal of that energy can be used more efficiently.

By anticipating the traffic flow you can keep your foot off the brake pedal for some distance. In other words, if a driver gets off the gas and coasts along on built-up momentum instead of running up to the stalled traffic at cruising speed and jamming on the brakes, he will save fuel. Coasting along on speed that already exists is free ground covered in the interests of energy conservation. The more distance you coast to the braking point, the more gas saved.

If the gear lever is slipped into neutral for the coast, you will go even farther, but coasting in neutral is against the law in some states.

This habit of coasting to a stop must be developed carefully, and the driver must stay alert to what is happening around him at all times. He must, of course, be prepared to stop at any time by using the brakes as hard as necessary.

The economy driver will learn to use his right foot once again for the brake pedal, in spite of the current fashion of left-foot braking in cars with automatic transmissions. The most important reason for using the right foot to brake is to avoid using the brake and gas pedal at the same time. The right-foot braking habit also eliminates using the brake pedal for a footrest, an unconscious error obvious to following drivers who see blinking taillights. Even the slightest pressure on the brake pedal causes some drag, while the lights use tiny bits of energy at each flash. If the driver learns to move his foot from the gas to the brake, a time lag in between will remind him to use the built-up speed before he touches the brakes.

It is possible to save fuel by discontinuing the use of the hand or parking brake. Use the parking gear for secure parking on an automatic transmission, and leave a stick-shift car in gear. The hand brake is actuated through a mechanical cable and this can hang up on release, causing a slight but very real drag. A dragging brake will cause any car to use more fuel to maintain a given speed. The use of the hand brake is not recommended at all in cold climates where slush may wet the brake; the whole mechanism can freeze when the car is parked overnight. It will take some running and brake kicking to dislodge the frozen shoe and cable, which in turn exacts a penalty in extra gasoline.

Developing a style of driving that keeps brake tapping to a minimum and uses brakes only for bona fide stops will go a long way toward improving the fuel economy of any type of car.

CITY DRIVING METHODS

Stop-and-go traffic is a big waste of gasoline for any driver. The full-scale traffic jam forces any car to use more gasoline than can be recovered no matter how carefully one drives. The main rule in city driving is to avoid traveling at peak traffic periods whenever possible and to stay away from roads and streets known to be congested. Following the tips on how to start, stop, and gauge traffic lights will bring your city driving economy up by several gallons in a week's worth of commuting.

Choose your urban routes with care. For the daily journey try several different routes to find the least traveled and congested. It is not uncommon to find a quicker route to work on ordinary streets than is available on a freeway, parkway, or other limited access road. Urban toll roads can be hard on gasoline consumption, too, if traffic stacks up at the tollbooths, inching forward to the gate. This sort of on and off action with the gas pedal wastes time and uses a good deal of gasoline for very little forward motion.

It should be simple to find the smoothest and easiest route for a regular trip, and to adjust driving speeds to keep the car rolling at all times. It will really save gasoline, and the weekly fill-up should be a gallon or so less than normal.

For the occasional drive into unfamiliar territory consult a map before plotting a course. Choose multiple lane roads where possible, or any street that appears to be a main road on the map; the traffic lights should work in your favor. Know where you are going by using the map before departure. You will save time as well as the gas that might be lost on repeated stops to ask directions, not to mention getting lost. If you are traveling alone, write the course on a slip of paper; it's easier than pulling over and referring to a map.

Conserve fuel by trying to anticipate every possible happening on the street ahead. Allow a comfortable distance between your car and the one in front. This is not only a good safety practice, but if the car you are following is on and off the brakes, you can keep a steady throttle without concern. Chances are you will arrive at the next stoplight right behind the brake artist ahead, and you will have burned much less gasoline getting there.

On limited access roads stay in one of the middle lanes to avoid the anger of the faster drivers in the left lane, and the traffic jousting of those entering and leaving the freeway in the right lane. On urban throughways, too, a generous distance between you and the next car will allow you to set your speed at 50 mph (or whatever is preferred) and stay on it with a steady foot. Should you need to slow for traffic you will have ample room to decelerate by coasting rather than by using the brakes.

On ordinary streets use the left traffic lane if there are left-turn lanes marked on the street. Otherwise use the right lane to avoid being hung up by those turning left, which often results in minutes of stationary idling.

Use the same technique at stop signs as for red lights—get off the gas pedal when you see the sign and coast to it. Make the full stop, then accelerate gently away from the intersection. Try to avoid making left turns against busy traffic. Sometimes it saves both time and fuel to go completely around the block in order to enter a parking lot from the right, instead of waiting out a herd of traffic to make the left turn.

Fuel can be saved with the proper approach to an intersection when traffic is flowing smoothly. The average motorist automatically slows down and often brakes close to an intersection even though the light is green and cars are moving along. This habit not only slows the flow of traffic, but causes fewer cars to get through the intersection with each change of the light. A steady speed is the best method of saving gas, and each time the foot is on and off the gas pedal the accelerator pump may be brought into play, squirting extra fuel through the carburetor. Furthermore, the loss of momentum requires more pedal pressure, and more fuel, to regain speed after crossing the intersection. Try to adjust your driving style to achieve a steady foot on the pedal, and eliminate unnecessary deceleration at intersections.

Cruising around the supermarket parking lot looking for a spot close to the door is a waste of gasoline. (The short walk might also provide some needed exercise.) Park as soon as possible, but better yet plan your shopping trips to save fuel. Combine the day's errands into one journey and you should be able to reduce fuel consumption by 20 to 30 percent. Many people make a number of short trips during the day, and the car stands for hours between runs. The engine becomes cold and uses additional fuel warming up to operating temperatures after each start. Someone who ferries children to school might consider arranging errands to dovetail with one of the day's two journeys, using the warm engine to get it all done in one trip. Finally, running to the market several times a week is a bad habit. Shopping can be planned, a route devised, and the week's marketing completed in one giant spree.

In general, city driving uses a good deal more gasoline for the miles covered than highway driving. However, using these tips will conserve fuel in the heart of any city for the careful driver.

BEAT THE SIGNALS FOR MILEAGE

Both time and gasoline seem wasted at a traffic signal. In congested city traffic one often has to wait through two or more complete cycles of a three-way signal before clearing an intersection, and this may take several minutes. After waiting out a long tedious signal the typical motorist is inclined to really push his engine to make up the time, and with that angry stab at the pedal another shot of gasoline is wasted.

Learning the elapsed time of traffic lights on a familiar route, such as the daily journey to school or work, can save you gasoline and indigestion. The average light is red for around 20 seconds, although the timing will vary. Busy intersections where there is greater traffic flow on one side of the intersection have the lights biased toward the main road. However, across the country the majority of traffic signals use a 20-second red cycle.

Study the signals where you regularly drive and establish a time span for each one. You need not have the distraction of timing with a stopwatch; count to yourself one thousand and one, one thousand and two, and so on in normal cadence. You will approximate a second each count. It will also tell you how many seconds have elapsed when the light changes. Learn the time on each intersection for both red and green phases, and check the amber light timing as well. Some amber lights are nearly instantaneous while others are on for several seconds, especially when a signal controls a multilane street.

The next step is to attempt to hit all the lights on the route while they are green, thereby saving gas. Anytime a complete stop can be avoided you will save the amount of gasoline it takes to get a car rolling from a position of rest. Determine the speed to travel in order to make all the green lights, slowing occasionally but never coming to a complete stop if possible. It will surprise you to find that this average speed is usually quite close to the posted speed limit—on most city streets it is 35 mph. The catch here is maintaining that average speed; in traffic it is almost impossible to drive a steady speed without darting in and out of lanes and using other less than courteous tactics. The trick is to learn to choose the fastest moving lane in order to arrive at nothing but green lights. Look ahead and start counting down the red light as soon as it goes from amber to red, then adjust your speed to arrive at the intersection just when it is due to go green again. You will have to contend with the unpredictable behavior of the drivers ahead of you, but with a little practice you will soon find just the right pattern, thereby hitting mostly green lights on a familiar route.

Along unfamiliar territory you might adopt some rules used by drivers who competed in the old Mobil Economy Run. The first rule was that highway lights are heavily in favor of the main road. In other words the side road gets about 10 seconds of green whereas the highway will get anywhere from 30 to 40 seconds of green with a 10-second red. Generally the main road will have a 4- or 5-second amber light as well, and half of that time can be added to the green for calculation. Often the side road will have a tripper-type signal, and the main road will have green lights all the time except when cross traffic is present. While traveling, then, develop the habit of watching the side roads on signal intersections and be prepared for a light change, usually a quick one, if cars are approaching on the access roads. Generally the light will change from 5 to 10 seconds after the car trips the signal, but the time will vary with the amount of traffic. Getting off the throttle when you see a car approach on a side road can allow you to coast slowly to the red light without using the brakes, arriving just as it turns green again.

Another rule of professional drivers is to use the amber cycle in an all-out effort to keep from coming to a complete stop. Use it with caution because no one should develop the habit of running red lights, but in many states if the car or part of it is in the intersection when the light goes red it is legal to continue through the intersection, even though the amber has changed to red for most of the distance.

Check local laws on this one.

In some areas traffic lights have a "Walk–Don't Walk" display along with conventional lights. When the flashing "Don't Walk" goes to a steady state, it tells the approaching driver that the light is about to turn red. Watching this display on the side road can provide a clue to the light you are approaching, and the steady "Don't Walk" on the side road tells you your light is about to go green. The sharp-eyed driver can catch the flash of amber on the side road for another clue of impending green on the main road.

There is one trick that can help make green lights in unfamiliar country. If the road is lined with utility poles, you have a natural measuring stick for the light change even if you have never been on that road before. It works this way: If the traffic light is green when you pass the last utility pole before the intersection, it will stay green or be just changing when you are midway through the intersection. It sounds like an old wive's tale but it has worked for many years for Mobil Economy Run drivers. If the light flashes to amber at or before the last pole, prepare to decelerate and brake gently for the red light.

Another trick for anticipating a light change is designed for intersections around a curve where signals are not visible. Watch the flow of oncoming traffic. If traffic is approaching you in any volume, it means the light is green and you will probably bag it red. If there is little or no traffic coming, the light is red and should go to green by the time you see it. Of course, you should be prepared to stop because not all light cycles work on schedule. However, except for three-way signals and multiple left-turn signal intersections, few lights are red longer than 30 seconds in any cycle.

When dealing with a long red light it is often more thrifty to shut off the engine than to sit and idle. Although it depends on engine size, most medium V-8s reach an equal gas consumption point for restart against idling at about 45 seconds. If you catch a long-cycle red light, turn off the key and slide the gear selector into neutral. All automatics will start in neutral, and it is an easy slide into the drive gear. Watch the cycle and when your turn is next (normally after the left-turn arrow appears in your direction of travel) start the car a few seconds before the anticipated green to be ready to move out with the traffic flow.

Beating signals is a fascinating game, and probably the most entertaining of all the methods of economy driving. It can turn a dull daily trip into a competitive diversion. Keep alert to seasonal changes in traffic flow and traffic light timing, and it won't be too long before your fuel economy will increase because the lights will all be green. And it is guaranteed to reduce frustration.

Winter Driving Is Special. Fuel economy will drop in cold and stormy weather. Even rain can hurt fuel mileage because a wet pavement with puddles increases the rolling resistance of tires. Snow and ice often call for the use of studded tires and chains to improve traction, and these also impose an economy penalty due to the extra friction between the tire and the road surface. In addition, a cold climate calls for heavy use of heaters and defrosters plus windshield wipers, and these all consume energy.

There are ways to save a little gasoline while driving in the winter. Be sure that car windows and doors seal well, minimizing the heat loss from the inside. Just as you would in your home, turn the heat down. In other words, become accustomed to a little less heat and try to reduce the use of the blower fan, which uses a considerable amount of electrical energy. Don't compromise safety but use the electric rear window defogger sparingly if you have one. The electric wires in the rear glass do an excellent job of cleaning the window but remember to turn the current off when the ice is gone. Electric power costs gasoline.

Steamy windows are a big safety hazard in winter driving, and they always seem to fog the most when the car first starts in the morning. Initially the engine is cold, and there is little heat for defrosting the windshield no matter how much fan air is blasted across the glass. Save fuel and leave the blower alone at first. To avoid fogged windows in cold weather roll down the car windows, at least the driver's window, before entering the car. Letting cold air in will help keep the inside fog-free. Unless ice has formed on the outside, keeping the inside temperature close to the outside temperature will alleviate the steamy window problem. Obviously, in sub-zero weather this technique will be so uncomfortable as to be impractical.

As mentioned above, moderate the use of traction aids because they cut rolling efficiency. Never use chains unless they are really necessary. Chains do dictate slower driving, and that helps, but driving on bare pavement with chains not only wastes gas, it tears up the chains and damages the pavement.

Inspect your car often in the winter to be certain unwanted

pounds of extra weight in the form of frozen slush are not stuck under the fenders. Kick, knock, or prod such deposits off. Avoid setting the parking brake because it can freeze in place and create drag.

Drive only when essential in nasty weather, consolidate short trips into one longer excursion, and plan to go when the weather is most favorable. Finally, accept the fact that cold weather will require more fuel, but with a little care fuel can still be saved.

COLD STARTING

As emission control systems have proliferated, it has become increasingly difficult to get the automobile started first thing in the morning. The colder the outside air the harder it is to get a late model automobile moving. Some people have returned to the old-time habit of firing up the car in the driveway and letting it run at fast idle while they finish their coffee. Although this ensures that the car won't stall in the middle of the street, it is very wasteful of gasoline.

Every car has cold start instructions in the owner's manual, and after a month or two with an automobile you should know how best to start it. Usually a late model car calls for one full depression of the gas pedal before the ignition is switched on. With the possible exception of very cold weather, the engine will catch if no more gas pedal pressure is applied when attempting to start. (Normally, pumping the gas pedal does more harm than good.) The automatic choke will come into play and the car will run in neutral or park at a fast idle. If the accelerator is tapped hard, the car most generally will come down a peg or two on the choke and the engine will die. This is what makes it so difficult to get out of the driveway on a cold morning. Remember, every time you have to crank the starter to refire the engine extra energy is being used and it takes gasoline to replace that energy.

For the car with a manual choke, a standard feature of many imports, a half choke setting is usually sufficient for a cold start unless the weather is very cold or very warm. Remember to push the choke in once the engine temperature starts to show on the gauge. Some imports have a red warning light that glows on the dashboard until the choke is completely out of action.

The most frugal way to start out with any car is to use the gas being burned with full choke to get down the road. On the standard

American car with automatic transmission this is not so simple because often the engine will die while the lever is being moved from park through reverse to drive. It is a matter of trial and error to find just the right throttle combination to keep the machine alive with a cold engine. Try to avoid giving it too much gas in reverse while backing out the driveway. A good stab on the gas pedal will bring the choke off its top setting, and as the car is slowed, stopped, and shifted into forward gear, the engine may quit. The ideal solution is to back in at night and be ready to go forward in the morning. Doing this, as well as starting the car in neutral instead of park, there will be less of a lag between starting and getting under way when the car is into drive.

When the engine does catch give it just enough pedal pressure to get moving, and if it starts to stutter, give it more gas gently. Get down the block while it is running on the choke. Although the rich mixture needed to start a cold engine is far from economical, you can at least cover some ground while all that gas is flowing. This is far better for saving fuel than letting the engine idle away gas in the driveway. After three or four blocks, unless the outside air is quite cold, tap the pedal to drop the choke to a leaner mixture. A manual choke may be pushed in almost completely.

The ideal morning route should have a mile or two of slow driving before getting on any throughway. The slower driving will ensure that the choke is truly off, the engine and transmission lubricants warm enough to be effective, and the tires warm, flat spots having been eliminated. At that point you are ready for highway speeds.

A rich gasoline mixture is required to start an engine in cold weather. The big tip for fuel savings is simply to keep it running and get a couple of miles out of that rich flow of gasoline. Know when to knock down the choke to avoid running extra miles on a partial choke that is no longer required.

MOUNTAIN DRIVING

Driving in hills and mountains will require more gasoline at the same speed than a car will consume on flatlands or in slightly rolling country. Driving an automobile up an inclined surface will take more energy to produce a given speed than running on a flat grade. Of course, every mountain road has a downhill side, and with practice a good driver can equal or better his level country mileage in the mountains.

The degree of the incline and the length of the hill make a big difference in the best method of climbing, and the type of drive train your car has is another big factor. For example, a two-ton sedan equipped with a very low axle ratio (often called an economy axle) and an automatic transmission may barely be able to climb a hill or achieve good fuel economy at low speeds. Low rpm is the villain here as the engine lugs, straining for power. In this case it is better, especially under 50 mph, to downshift into a lower gear and raise the rpm to a more efficient rate. Although this sounds completely foreign to the concept of fuel economy, it has been proven during factory testing.

The same idea also works well for low horsepower small cars. It is not particularly economical to lug a 70 hp one-ton sedan up a steep grade in fourth gear. It will save fuel, not to mention mechanical repairs, to downshift to a more comfortable gear. Unfortunately, the 55 mph speed limit makes impossible the old ploy of getting a running start for a steep hill.

A medium-sized or standard V-8 automobile has no good rule of thumb for efficient hill climbing. The variations of engine/transmission combinations currently in use make it impossible to find a simple answer to the problem. In most cases the best method for climbing a long steep grade involves a steady foot on the gas pedal. A gradual increase in pedal pressure while on the grade will hold the speed between 45 and 50 mph. The car will use some extra fuel, but a light foot and a steady increase in pedal pressure will increase the fuel flow slowly, and none of the gasoline will be wasted in poor combustion.

Coasting on the downside of a hill will save a lot of gasoline, but it must be practiced in a reasonable manner. Turning off the key would provide a completely free ride, but it is dangerous and illegal in most states. The next best thing is coasting in neutral but it, too, may be against the law. Coasting in top gear is the closest thing to actual coasting in neutral, and it is perfectly safe.

Coasting requires that the driver be prepared to restore power to the drive train at any moment. Never attempt to coast with the engine off in a car equipped with an automatic transmission. The transmission fluid pump may stop, resulting in damage. Vehicles with power steering and brakes are very difficult to handle without these power systems operating; the average driver would quickly find himself in serious trouble should he attempt to save fuel by using his car as a coaster wagon.

Sustained periods of mountain driving will serve to teach the best methods of getting good economy. High altitudes rob horsepower, so holding uphill speeds and passing on the upgrade become chores even under full throttle. Engines tend to overheat at high altitudes unless they are tuned for the elevation. Anyone planning to spend extended time in the mountains should have the timing and the carburetor reset for high altitude. The engine will perform better and fuel economy will be improved.

HEAVY LOADS VERSUS GOOD MILEAGE

It is a fairly obvious fact that the heavier the vehicle the more fuel it takes to move it. There are some other variables involved, most notably aerodynamic shape, but a two-ton passenger car normally uses about 40 percent more gasoline per mile than a one-ton subcompact. The heavyweight sedan ordinarily has an engine with three or four times the displacement of the small model.

We know that late model cars do not achieve the economy of similar models from the 1960s. Part of this loss can be traced to a weight gain from safety equipment mandated by law. There is nothing the consumer can do about the extra weight. In the interests of efficiency, the consumer can juggle the weight he adds to the car and avoid carrying any unnecessary cargo.

The proper loading of baggage can reduce your fuel consumption on the highway. Roof racks are the worst gas user. They protrude in the airstream, enlarge the frontal area of the vehicle, and often cause top-heavy instability in high winds. Stow all luggage *inside* the car when trying to get the best economy possible. A heavy load in the trunk is poor for fuel economy because of the problem of unequal weight distribution. Weight in the rear of the car raises the front, increases drag, and reduces the aerodynamic efficiency of the vehicle. The ideal place to stow baggage is in the center of the car, perhaps on the floor of the back seat or as far forward in the trunk as possible. Keep the weight as near the center as possible so that the vehicle stands level on the road.

PASSING TIPS

In ordinary driving passing another car usually requires a bit more speed with an attendant loss in fuel economy. It is possible to use less fuel in passing but it takes a different driving style than most of us are accustomed to using.

On a multilane highway, passing traffic with a steady throttle setting is no real problem, and it is more than simple on divided highways. With the new lower speed limits in effect the opportunities for passing are not as great, but we should know what to do, slow or fast. The best technique is to estimate the traffic flow and to plan the move so that you can pull out, pass, and get back into the right lane without changing speed or blocking faster traffic. Look ahead and use the mirrors. It may be necessary to decrease speed a little to avoid jamming up a lane; if so work down to the proper speed gradually before starting to pass. Should passing require increasing your speed, watch for a downgrade where the acceleration will cost less fuel than on the flat or—even worse—uphill.

Passing on a two-lane highway and keeping an economical constant throttle setting is difficult. If traffic is heavy, it is virtually impossible, because one needs plenty of room and no oncoming traffic to pass another car safely while traveling only a mile or two faster than the car being passed.

Usually it is necessary to abandon economy driving tactics when passing on a two-lane road in order to get the car around and safely back in lane. However, it should not be necessary to floorboard the throttle either. Dropping back a bit more than normal from the car ahead allows more space to work up to the passing speed with a gentle acceleration. Be sure that the road ahead is clear of oncoming traffic, and be prepared to duck back in behind the car you are attempting to pass, or to give the throttle a healthy stab to complete the pass.

Passing in city traffic is a matter of staying steady on the gas pedal and anticipating the movement of traffic ahead. The driver must always be aware of what is going on to his rear, especially on multilane streets. Indiscriminate darting from lane to lane on urban roads may help maintain a constant speed, but the habit is dangerous and may result in a traffic citation. It is more reasonable to adjust the driving speed to the traffic flow. Estimating speed in traffic is an acquired skill and such judgment is a big help in knowing when to pass in city driving without undue acceleration.

WHICH GEAR—AUTOMATIC

The American-made automatic transmission is a marvel of convenience, but it is a bit less efficient in the transfer of power from engine to wheels than is the standard stick-shift transmission. Most automatics replace the manual clutch with a liquid coupling system that eliminates the need for a clutch pedal. Today, the three-speed automatic is almost standard in domestic cars, and knowing how to use the gears can save gasoline in many driving situations.

Downshifting an automatic for better economy while climbing hills has been dealt with in the section on mountain driving. Getting into drive or high gear as quickly as possible is quite economical, as outlined in the section on starting off. Beyond these habits the automatic will get the best mileage in high gear in nearly all situations because the lower rpm in drive is more frugal than running in second gear at the same speed at higher rpm.

Moving the gear selector into neutral at stoplights will eliminate the tendency to creep but the fuel saving, if any, will be negligible. Getting out of gear may be more economical if one is coasting to a stop. Once in drive let the car creep away from lights before touching the gas pedal. Although this start is not free, an idling engine will propel a car at about 5 mph, enough to get you rolling before pressing the accelerator.

WHICH GEAR—STICK SHIFT

Not so many years ago the old reliable column shift, the three-speed manual transmission, was a staple in every showroom. Today it is seldom available except as optional equipment, although it remains popular with drivers of fleet vehicles such as taxicabs. The stick-shift car will achieve its best fuel economy during normal driving in third or high gear. Most of the time, at least on level terrain, the driver can forget about first gear, using second and third, and handling most traffic situations in third.

Manual transmission cars fitted with a four-speed gearbox have it generally for one of two reasons: a low powered engine needs four forward speeds to provide good overall performance, while big cars with four speeds are usually sports cars or high performance sedans, like the Corvette and the Jaguar. In any case, fourth gear will produce the lower rpm for a given speed, and in turn will provide the best gas mileage for most driving situations. One exception, as noted earlier, is in climbing steep hills with a low horsepower car, in which case high gear simply cannot be used. Another exception is in urban driving, around 30 to 35 mph, with a small four-cylinder engine; the best performance will be in third gear. Rpm will be higher than in fourth gear, but the engine will not be straining from lack of power, and the fuel economy, overall, will be better.

A number of imported cars are fitted with five-speed manual transmissions. The sports models of Porsche, Alfa Romeo, Datsun, and Toyota, to name just four, are so equipped. In most cases the fifth gear acts much like an overdrive. It is valuable on the open road for economy but is relatively useless at speeds under 60 mph. As a result, a 55 mph speed limit means that such cars must be driven in fourth on the open road to avoid lugging the engine. As an example, a Porsche 911 returns around 28 mpg traveling between 55 and 60, but returns about 32 mpg as the speed is boosted to 70. The difference is that the Porsche is lugging in fifth gear at 55, but performs very well at 70 in fifth.

Lugging the engine is abusive and may result in far more damage than is justified by the small saving of fuel. Riding the clutch is another bad habit. Putting the gear lever in neutral while stopped is a good habit, and when one's foot is off the clutch it is saved from unnecessary wear. Break the habit of downshifting when coming to a stop. It is very sporting but it costs gasoline each time you do it. Use your regular brakes and do not depend on engine braking, in keeping with the idea of coasting to a stop wherever it is safe to do so.

Saving an ounce of gas here and there may seem ineffective but those ounces will add up into significant amounts, especially in city driving. Imagine saving an ounce at each traffic light. Eight stops per day would be a water glass full of gasoline. Every four days it would be a quart, and close to two gallons per month.

FEATHER-FOOT TECHNIQUES

"Feather-footing" is the best way to get good gas mileage from any automobile. Imagine that your foot weighs no more than a feather, or suppose that a fresh egg lies between your foot and the pedal. A gentle touch on the accelerator saves more gasoline than any other single method of economy driving. The motorist who is dedicated to getting the most miles per gallon will probably develop a completely new style of driving. He may even change his car.

No matter how well designed the drive train of a particular vehicle, the careful driver can increase his fuel savings by about 30 percent. Easing away from stops, coasting on deceleration, driving at a steady pace on the open road, never jamming the pedal to the floor—these are the feather-foot techniques that make the big difference.

However, feather-footing is an acquired skill, and definitely foreign to almost every driver on the road. It takes a lot of concentration to maintain a steady highway pedal pressure, and even more work to outguess the traffic lights.

Driver comfort is critical to success. Before starting the engine the driver should be sure that all the necessities of life on the road are within reach. Sunglasses, cigarettes, and snacks should be within easy reach without moving the lower half of the body or the throttle foot.

The good feather-foot driver makes a game of getting where he is going on the least amount of fuel. It takes a strong desire to maintain a light throttle pressure on the highway and when crawling away from traffic stops. Initially the stop and go driving is tough to handle because the slower starting makes the driver feel as though he is foolishly blocking traffic. For daily driving it is best to find a happy medium between the traditional full power dash and the pure economy takeoff. This will make it possible to keep up with surrounding traffic and still save a good bit of gasoline.

THINKING AHEAD

Anticipating what might happen on the road ahead is one good method of saving fuel. The driver who is ready for the sudden stop, the unexpected buildup of traffic, and similar problems will be able to handle his car without a lot of unnecessary work on the gas or brake pedals.

Smooth driving produces the best fuel economy, and thinking ahead enhances your ability to drive smoothly. A 12-second visual search pattern has been recommended by experts to give you time to scan the road ahead and anticipate trouble before it happens. That is to say, a smooth stop can be made by the driver who tries to observe what might happen on the road 12 seconds ahead.

Thinking ahead can be an entertaining game. Watch a particular vehicle and guess what its next move will be; it can break the monotony of a dull ride. On hills, thinking ahead can help prevent being stuck behind an underpowered vehicle as it crawls up the grade. Decelerating to match a 20 mph truck on a long grade is very bad for fuel economy.

There are patterns to the flow of traffic on any road. Some main routes have regular trouble spots, while others are noted for being the quick way to go. In Los Angeles, for example, some freeways traditionally are faster moving than others, and it seems that these are the ones that carry less commercial traffic. Check the traffic flow by thinking and looking ahead, and when a block is spotted there will be room and time to seek an alternate route.

On the open road the habit of thinking ahead will keep a driver more alert. Most motorists, isolated in living-room-style comfort, tend to be too complacent. Thinking ahead will save gas and provide an extra margin of safety as well.

ECONOMY DRIVING TECHNIQUES AND WHEN TO USE THEM

Economy driving in professional competition has produced a bag of tricks that can be of value to the layman seeking better mileage from his car.

A little extra air in the tires will decrease rolling resistance and result in a small but definite increase in fuel economy. Gross overinflation has been tried, but it causes poor car handling and rapid tire wear. For the average highway trip inflating the tire pressure to the maximum recommended for a heavy load is all the extra air needed. At all speeds keep the tires matched in air pressure (front same, rear same). In urban driving extra air pressure will not greatly increase gas mileage, and it will make the ride less pleasant.

Certain interior ventilation methods can mean a loss in fuel economy. Driving with the windows down increases aerodynamic drag. Economy drivers in competition will struggle with a nonair-conditioned, closed window environment rather than turn on the air conditioner or open a window. In cold weather using the heater will also adversely affect fuel economy. Admittedly, no one in ordinary driving would be expected to live such a spartan existence. In mild weather, however, it makes sense to keep the windows closed at highway speeds.

Some imports have rear window extractor vents, so that opening air vents in front will provide ventilation without a fan. The next best thing is to use the front vents for fresh air intake and to crack one rear window about an inch to provide an exhaust vent.

Moderating the use of electrical accessories is another economy tip from competition drivers. The current fad of driving around in daylight with the headlights on burns extra fuel for no real reason. Only on a two-lane country road does the daylight use of lights make safety sense. Turn the lights off when not required, and if in doubt about when they are required, consult an almanac for official sunrise/sunset times and use the headlights accordingly.

Smokers should use matches instead of the car's lighter. The heating element is a big user of electrical energy and should be avoided if one is serious about saving that extra ounce of gas. Along these lines, avoid the use of coffeepots, trouble lights, vacuum cleaners, and other accessories that plug into the cigarette lighter. Use a flashlight, and carry hot beverages in a vacuum bottle.

Keep the car clean to avoid carrying around a lot of excess weight in dirt. Excess baggage is better left at home. Don't keep that bowling ball or those clubs in the car unless you are driving to the alleys or the golf course. We have heard of fanatics who claim that a good coat of wax will enhance a car's slippery shape in the highway airstream. That one is hard to buy but the owner who keeps his car clean tends to be the owner who inspects his car regularly for wear and has frequent, scheduled service work performed. In other words, a car in good mechanical condition will always deliver better performance than a neglected one.

SHORT TRIP TIPS

Energy conservation and short trips are not truly compatible, but there are ways to make short runs and save fuel, and the best method is to plan ahead. Follow our earlier tips on shopping and consolidate the short runs into one full day's journey, visiting all points in one big swing through the area.

The weekend away from home for the average driver usually constitutes a short trip—100 to 150 miles. The best gas-saving plan for this type of weekend is to estimate the hours to be spent on the highway. Remember to allow more travel time for lower speed limits. The best bet is to try avoiding the big traffic tie-ups during the Friday night exodus and the Sunday night return crush. To save the gas wasted in heavy congestion arrange to leave for the weekend either before or after peak traffic periods. If you cannot get away at noon on Friday, arrange to leave after dinner. Leave your resort or camping area a little early on Sunday or later than normal. Your on-road time and fuel consumption will both be reduced by avoiding the heavy traffic hours.

Consider the 10-gallon weekend, an idea put forth by the Recreational Vehicle Association. It may mean checking the map for weekend spots closer to home, and because these closer spots may be more crowded it will mean more advance planning.

There is nothing wrong with short trips, but the economy-minded driver will find they require some changes of habits before they make reasonable motoring sense.

ECONOMY TRIP ROUTE PLANNING

Planning a trip with fuel economy in mind can be a challenge. A vacation journey need not take as much fuel as in previous years, and simple planning can eliminate long hours on the road. Instead of planning on two weeks at a resort 600 miles away, try driving a total of 600 miles on the road, stopping for a few days in several different areas for variety. It is going to take longer to reach any destination because of the 55 mph limit, so be sure and plan lunch and overnight stops with that in mind.

One good way to save fuel on a trip is to plan a circular route. In other words, leave home on one highway and return on another. Each road should offer a variety of side trips, while the total mileage need not be long even though the trip is an honest vacation.

Business trips can be planned for economy, too. List the places to be visited, then make a grid on a map with lines connecting each town. It may look like a spider web but a close examination will reveal how to cover all the points and travel the fewest miles. It may involve crisscross travel but it is amazing how many miles can be eliminated with careful routing and the consolidation of several short runs into one longer business journey.

PICK YOUR DRIVING WEATHER

Weather has its effect on fuel economy. As noted in the discussion on winter driving, the stormier the weather the poorer the gas mileage—a fact that relates directly to the amount of resistance as the car rolls along the road.

Snow is a real enemy because the wheels must push it aside, and this means substantial rolling resistance. It takes a lot more energy to produce the same speed on a snowy roadway than on dry pavement. Ice does not offer heavy rolling resistance but it does call for some fancy footwork on the gas pedal, as well as downshifting and braking—all bad for fuel economy. Snow tires and chains further increase rolling resistance and take more gasoline than standard tires. Also, snow and ice call for use of the heater and defroster—a further energy drain.

Rain and partially flooded streets increase rolling resistance, and windshield wipers take a bit of energy. Lower speed in precipitation is a sort of bonus but the increase in rolling resistance usually offsets it. Poor weather makes the economical operation of a car difficult, no matter how you look at it.

Perhaps the biggest villain is wind. Driving into a head wind will eat up extra gallons of gas, and reducing highway speed does not compensate for a stiff head wind. On the other hand, a tail wind will greatly increase gas mileage. On a flat road the average V-8 standard size car will lose 1 to 2 mpg in a medium force head wind. At the same road speed the car will pick up about 2 mpg with an equally strong tail wind. The bigger the car the more gain from a tail wind, while almost any shape suffers a loss of fuel economy from a head wind.

It is impossible to pick your weather in daily driving but it may be possible to detour around bad weather on a vacation. Avoid planning motoring vacations during the bad weather season unless you are headed for a ski resort. Use snow tires and chains only when needed, and remove the chains when on clear pavement.

Winds are fickle but they seldom last long. A reduced speed in the face of a big wind helps, but it might be a good time to stop for an hour on the chance that the wind will subside.

Preplanning is the best method of avoiding bad weather on a trip. Determine what kind of weather conditions can be expected for a certain area and a certain time of year. Sometimes a better weather route can be found; sometimes it may be necessary to endure the worst. Picking the right time of year is usually the best way to pick weather that will help deliver the best fuel economy.

15

The Right Car, The Right Equipment

MATCH YOUR VEHICLE TO YOUR NEEDS

The day of the businessman who buys the largest car he can afford as an expression of his affluence and as a means to impress his associates is all but over. Most of us have become more practical. The smart motorist matches his vehicle to his driving needs, and saves fuel by not buying more weight and power than he really needs.

How can you make a happy match between your needs and the car you buy? Common sense is part of it; all too frequently expensive automobiles are purchased on impulse. The color matches a favorite suit, or the upholstery smells new, or the stereo sounds romantic, or some equally frivolous reason.

Every motorist knows that it is impractical to expect 30 miles per gallon from anything but a very small automobile. But he should know that he need not accept such grim numbers as 7 or 8 mpg, although special circumstances such as a large vehicle towing a heavy trailer may make very low fuel economy unavoidable. The trick is to strike a happy medium, namely the best mileage possible from a well-designed vehicle, driven and maintained by a conscientious and knowledgeable motorist.

Let's cite some practical examples. In the months since gasoline came into short supply small cars—both domestic and imports—have become so popular that dealers have difficulty keeping them in stock. Yet small cars—be it Pinto, Vega, Datsun, Toyota, or VW—are not for everyone.

A family with four growing children, for example, is going to have a rough time fitting all hands into a so-called six-passenger subcompact for anything but the shortest journey. If that family cannot plan on leaving someone home on long trips, they should look for a slightly larger car. There are intermediate size vehicles that do almost the same job as a full-size sedan, and use less fuel in the process.

A retired couple might do just fine with a very small car and be very happy with the great fuel economy—unless they also enjoy traveling with a trailer, in which case they will need a car powerful enough to haul that trailer. And the proper size vehicle might turn out to be a full-size sedan or even a light truck.

What we are saying is simply this: Don't buy a car too small for everyday use just to save fuel. On the other hand, don't buy the largest thing in the showroom, unless you can demonstrate to yourself that the space will be required on a regular basis.

You may find it more economical to buy a compact for daily use and rent a large sedan or station wagon on those occasions when more room is needed. Also, renting a small car with good fuel economy may be worth thinking about for the motorist who maintains a large car for business but does not need all its space and weight for an occasional vacation trip.

In truth, a typical suburban household, one comprised of a married couple with two or three school-age children, does have a dilemma. What is the best way to handle the large-scale transportation chores (Little League, Scouts, etc.) in a community where the automobile is essential? Many families have found that two cars work best —a big sedan or station wagon for local driving with the full crew of youngsters aboard and a much smaller car for the economy required in daily commuting.

There is no simple answer. If there were the world's auto makers could fire all their planners and researchers and concentrate on building the models they would know to be the best for all. In this country we have such a variety of driving conditions and personal requirements that the universal car just cannot happen. The motorist must consider his needs, survey the market to see which vehicle is best for him, and then drive and maintain it to the best of his ability. If he works hard at the last two, he will have the satisfaction of knowing that he is saving as much energy as possible.

CHOOSING A POWER TRAIN FOR ECONOMY

The object of the economy game is to get the most performance on the road with the least work from an engine. The efficient transmission of power from engine to rear wheels is a complicated problem involving the type of transmission and its gear ratios and the rear axle and its final drive ratio. Engine size and vehicle weight are other factors that enter the picture.

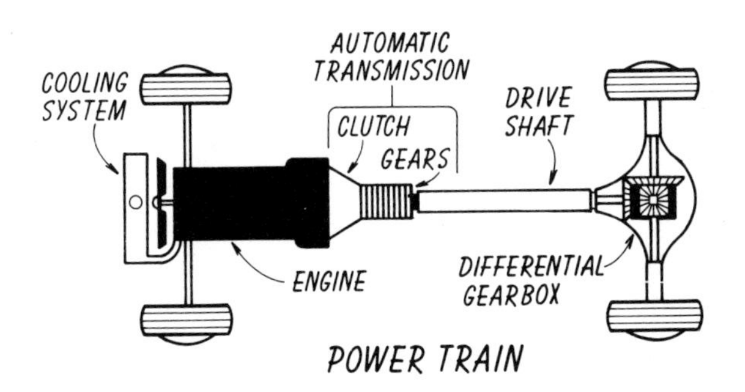

POWER TRAIN

What we must consider, however, is the choice available to the motorist. How can he make an intelligent selection of engine, transmission, and rear axle ratio that will suit his needs and deliver fuel economy? It is not as difficult as it sounds, even for those of us who are nearly helpless when it comes to the technical aspects of what makes the wheels go round. Remember a few simple rules and you will be able to astound your friends and, hopefully, make an intelligent choice of power train in your next auto purchase.

Rarely will you, as a customer shopping for an automobile, be offered a choice of power train. Salesmen are human; they want to sell the merchandise on the floor rather than have a customer submit a special order that may take weeks to deliver. Customers, too, hate to wait while their car is assembled according to specifications. However, the few weeks of waiting can pay off in terms of overall satisfaction and, more specifically, in measurable savings of operating costs.

Beginning up front, we will pick an engine. Although the variety of engine sizes and horsepower ratings has narrowed in the last half-dozen years, there remains a reasonable selection from which to choose. Full-size cars offer the greatest variety; with the small sub-compacts frequently there is no choice at all. In general terms the engine with the smallest displacement should provide the most miles per gallon. The exception might be a six-cylinder power plant in a large sedan, in which case the six would have to work so much harder than a medium-sized V-8 that not only would acceleration and general performance be unsatisfactory, but mileage too would be no better than with the V-8. Many car makers have recognized this fact and no longer offer the six in large sedans.

When selecting a transmission, be advised that the old-fashioned stick shift is the one that will deliver the best fuel economy. The primary reason is that in high gear there is no power loss as there is with the fluid coupling of an automatic transmission. The exception occurs when the stick-shift driver spends more time than necessary in the lower gears. Unfortunately, it is not always possible to order a stick shift, and in any case many of today's drivers have become so accustomed to automatics that they are unwilling to go back to shifting.

The third consideration is the rear axle. Rear axle ratios are expressed numerically. For example, 3:1 would mean that for every three turns of the automobile's drive shaft in high gear, a differential with this ratio would cause the rear axle and rear wheels to make one revolution. As a practical matter, axle ratios vary from about 2.5 to 4.5:1, depending on the vehicle's power train and its intended purpose. For best economy the general rule is to pick the lowest numerical ratio, which means that the engine will be turning at a slower speed than it would with a higher numerical ratio.

The problem with picking a very low axle ratio for the best fuel economy is that acceleration suffers. In some cases performance may become so poor that starting off from a traffic light is an agonizing process. Cruising in high gear at moderate speeds produces great mile-

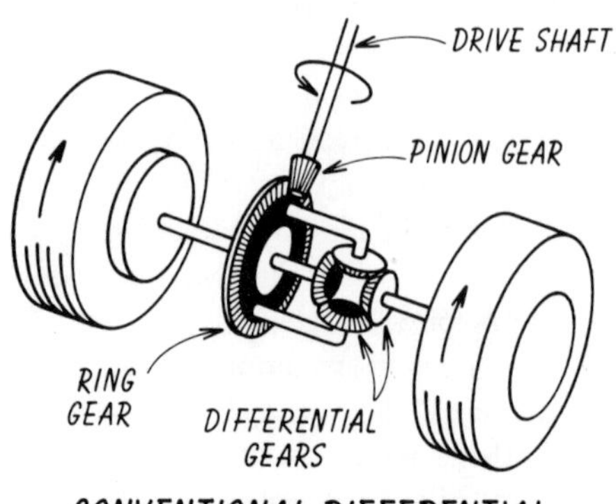

CONVENTIONAL DIFFERENTIAL

age, but passing acceleration is almost nonexistent. Fortunately, car makers are aware of these problems and tend to limit the optional ratios to those within the realm of practicality.

At the point of sitting down with the salesman and ordering the power train, ask which rear axle ratio is standard and optional. The salesman may not know the actual numbers but his book may list such options as "performance" or "trailering" ratios. Unless you are seeking an uneconomical, high acceleration performance, or plan to tow a trailer, avoid both these ratios for they are higher numerically than the standard one. The car will feel peppier, with better low-speed acceleration, but gas mileage will suffer. Look for and order the "economy" (or similar named) axle ratio. It is so named because that is what it is for. Expect to see more economy axle ratios as both motorists and factories place increasing emphasis on better gas mileage.

As examples of what you may find, here is the way Chevrolet listed its power trains for the 1974 full-sized automobiles, the Bel Air, Impala, and Caprice Classic. (In past years the Impala has been the world's biggest-selling model.) There are two 350 cubic-inch engines available for the Bel Air and the Impala, one with a two-barrel carburetor and one with a four-barrel unit. The Caprice Classic has a standard 400 cubic-inch engine with a two-barrel carburetor, which is optional for the other two models. Also available is a 400 cubic-inch four-barrel combination and even a 454 cubic-inch engine, optional for all models. It is interesting to note that Chevrolet no longer offers the six-cylinder engine in its full-sized cars because the cars have become so heavy that this engine is inadequate.

In the interests of economy we can rule out the 454 cubic-inch engine at once as overkill. It offers great performance, and is quite valuable for such chores as trailer towing, but only with the most delicate handling will it approach the economy of the smaller engines.

There is no choice of transmission; a three-speed automatic is standard. Stick shifts are still available in the smaller Chevrolet models, but no longer for full-sized autmobies.

There is a choice of axles, but each choice is coded trailer or performance. The best bet in this example is to stay with the standard axle. In the case of the 350 cubic-inch engines, the standard axle ratio is 3.08:1; the 400 and the 454 cubic-inch engine receive a 2.73 axle. The 2.73 axle tends to offset the poorer economy of the larger engines.

It isn't difficult to make a correct choice of the most economical power train, but it can readily be seen that a bad choice could be disastrous in terms of fuel economy.

POWER ACCESSORIES AND WHEN TO USE THEM

Any auto engineer asked to design an automobile for maximum economy would automatically eliminate every power accessory except those absolutely essential for the operation of the vehicle. Gone would be power steering, power brakes, air conditioning, electric windows, heater, defroster, cigarette lighter, radio, power seats, power door locks, and electric windshield wipers and washer. The only trouble with such a car is that no one would buy it. Chances are you couldn't even give it away, and it wouldn't even be legal because a heater/defroster is required by federal law.

The above is an extreme situation, an impractical and silly way to approach car design. Yet, every item mentioned requires energy to operate, and energy is gasoline. Some of the items take very small amounts or are used for such brief periods that the fuel required is of little consequence. The fuel requirements of every power device can be measured with a highly sensitive flow meter, but most of us would be reluctant to give up the comfort and convenience of such items as power brakes and power steering, each of which is virtually a necessary safety feature on a large automobile.

We cannot eliminate all the power equipment on a modern automobile, so let us analyze which ones are bad enough to rate a recommendation of "use sparingly."

The largest single offender is the air conditioner. It is very easy to see that the air conditioner takes power. Drive at a steady pace on level ground and turn it on. There is a noticeable change in the car's speed as the compressor cuts in. Another test would be to drive a set course with and without air conditioning, checking fuel consumption. It is not necessary to waste fuel in that kind of research; be advised that air conditioning will cost from 2 to 4 miles per gallon in urban and low-speed traffic, possibly more depending on engine size. A large engine in the 400 cubic-inch range delivering 12 miles per gallon may lose a couple of miles per gallon when the air conditioner is operating. Installing the same system on a small car with a 140 cubic-inch engine, for example, will probably cost 4 mpg or more. The power required varies. On a very hot day at low speed, when cool air is most needed, fuel economy is hurt the most. At high speed, say an illegal 80 mph, there is almost no penalty. The only answer for the motorist serious about saving fuel is to use the air conditioner sparingly.

The energy consumed by the heater is much less than that required for air conditioning. The main electrical drain is caused by the blower. The air conditioner also uses a blower, but it represents only a small part of its total energy consumption, the big drain for the air conditioner coming from its compressor.

Professional drivers in economy competition will go to great lengths to avoid using any power accessory, but they are worried about the most minute use of fuel. As a practical matter we should be economy-minded, but it would be silly to refrain from using a transistorized radio with very low current drain. The secret is to use common sense, to be aware of which accessories take the most energy, and to drive accordingly.

TIRES THAT ADD MILES PER GALLON

Tires are the one item over which every motorist has control, and it is a well-established fact that one type of tire will provide better fuel economy than most others. That tire is the radial, and it has other advantages such as good tread life. Many tire makers offer a 40,000 mile warranty.

There are three types of tires in general use, and it may be helpful to discuss them in order to better understand why a radial is beneficial. First is the bias-ply, a conventional tire with two or more plies of fabric that cross the center line of the tire at an angle. These crisscross layers provide plenty of flexibility for a comfortable ride, but this same flexing action extends into the tread area, resulting in a distortion that causes tread wear and reduces traction.

Stepping up in price and quality we find the bias-belted tire, which is a bias-ply inner shell to which supporting belts of fabric or some variety of man-made fiber are added around the circumference of the inner plies.

Finally there is the radial tire in which the body plies run radially from bead to bead. In addition, belts circle the tire to help stabilize the tread. Radials have sidewalls that are quite flexible (sometimes they appear to need air) and which cushion shock well, while the belts add puncture protection in the tread areas and, most important in terms of fuel economy, keep the tread flat against the pavement—for optimum wear and traction. Radial belts may be fabric but in many

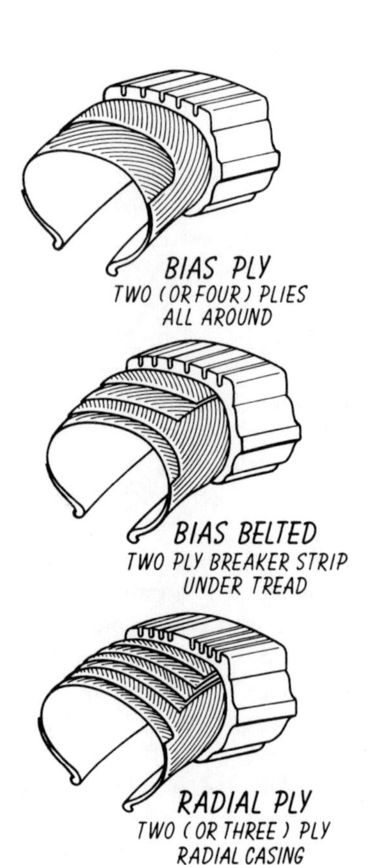
BIAS PLY
TWO (OR FOUR) PLIES
ALL AROUND
BIAS BELTED
TWO PLY BREAKER STRIP
UNDER TREAD
RADIAL PLY
TWO (OR THREE) PLY
RADIAL CASING

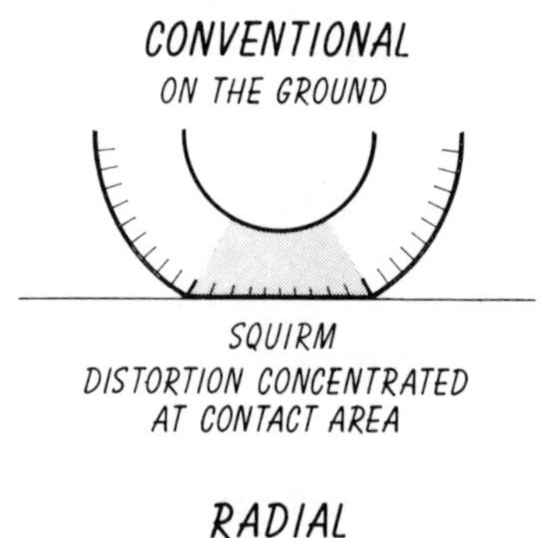
CONVENTIONAL
ON THE GROUND
SQUIRM
DISTORTION CONCENTRATED
AT CONTACT AREA

RADIAL
ON THE GROUND

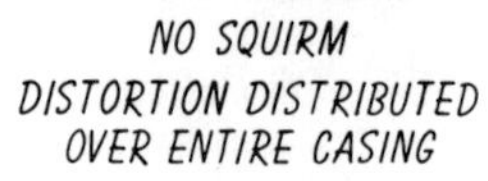
NO SQUIRM
DISTORTION DISTRIBUTED
OVER ENTIRE CASING

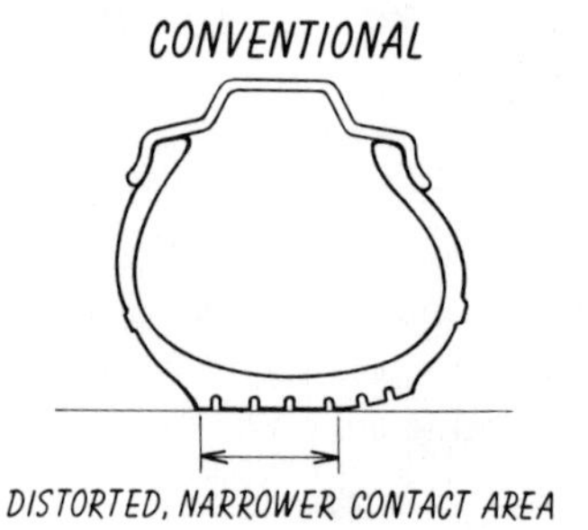
CONVENTIONAL
DISTORTED, NARROWER CONTACT AREA

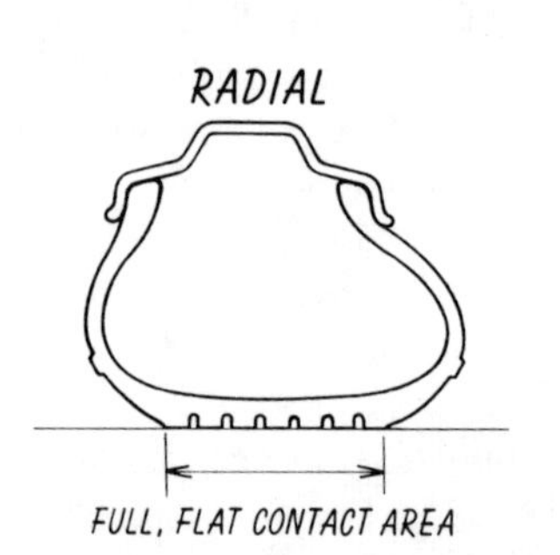
RADIAL
FULL, FLAT CONTACT AREA

cases they are metal. One major tire maker, Goodyear, says that their steel-belted radials have from 15 to 22 metal cords per inch, with each cord containing from 5 to 10 strands of metal. The metal cord is high carbon steel in which a 1-inch strip has a breaking strength of more than 2,500 pounds.

A great many tests have been run by tire makers, auto companies, and some government agencies, all in an effort to determine which tire is best. Everyone concludes that radials improve gas mileage, but the best test we have seen is one from Goodyear. They tested two 1973 Chrysler New Yorkers, equally loaded with the equivalent of two persons in front, one in the rear, and 100 pounds in the trunk. The tires tested were their own HR78-15 radials and a set of H78-15 bias-belted tires.

The cars were driven 24 hours a day at speeds from 35 to 70 mph for 19,000 miles—two thirds of this distance on interstate highways and one-third on rural secondary roads. Tires were switched from one car to the other every day to eliminate variations between vehicles, and drivers switched cars once a week.

Overall the radials delivered the best mileage during the test. Curiously, the biggest improvement occurred during the first 2,000 miles, and both tire types improved in gas mileage as the test continued. There is no explanation as to why fuel miles per gallon got better.

Early in the game—the first 2,000 miles—the radials produced about 7 percent better fuel economy, while at 12,000 miles the advan-

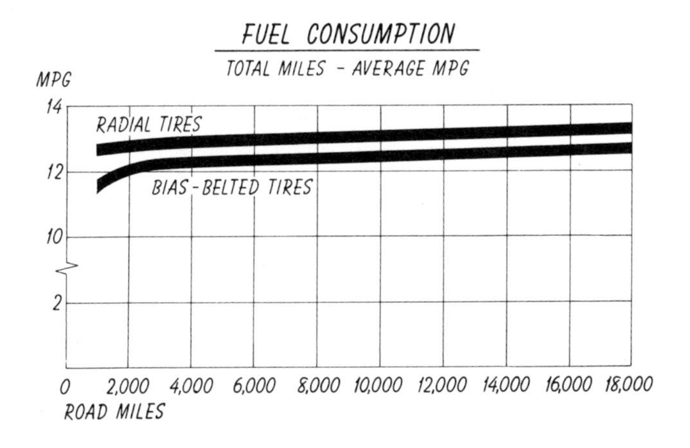

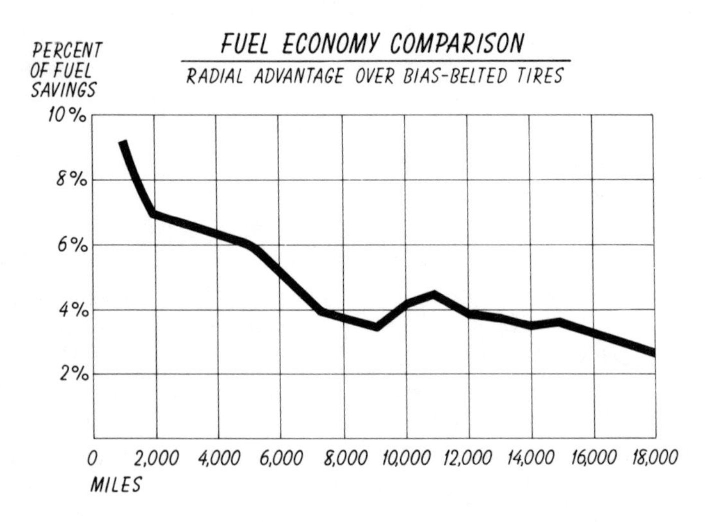

tage of the radials decreased to about 4 percent. Consequently, Goodyear claims an average annual fuel savings of 3 to 5 percent for radials over an extended period and this figure is compatible with other tests.

The disadvantage of radials is that they are more expensive than conventional tires, and they tend to change an autmobile's handling characteristics. Some drivers like what happens, others do not care for the additional road noise that occurs on some cars and the harsher ride. Before installing them the motorist should check the tire maker's applicability chart to be certain that his model of automobile is recommended for radials; not every car is a candidate. It should be noted that those U.S. auto makers who offer radials as standard or optional tires state that they "tune" the suspension for radials.

In brief, we will recommend radials as a fuel-saving idea that does work and that can be adopted for nearly every automobile.

EMISSION CONTROLS VERSUS MILEAGE

There is no question that emission control systems have taken their toll in reduced fuel economy, especially since 1970. A definite reduction measured in miles per gallon is impossible to quote because of the enormous variation in cars and engines among the world's car fleet.

It is a fact, however, that if every automobile in the United States equipped with emission controls could have them disconnected and have their engines tuned accordingly, overall fuel economy would improve about 20 to 25 percent. This is not to say that we condone the disabling or elimination of emission controls. Clean air is vital to our health, and we know that automobiles in urban areas can be blamed for the overwhelming bulk of air pollution.

Understanding the causes of today's relatively poor fuel economy does not relieve the motorist of the responsibility to do much better in conserving energy, but knowing the reasons why can be very helpful.

It really began to hurt when compression ratios were lowered in wholesale fashion. Reducing compression lowers engine efficiency, obviously bad for fuel economy, but it also permits the engine to burn gasoline with a lower octane rating, that is, gasoline with a reduced lead content. Now that we understand that the combustion products of leaded gasoline can be bad for people, we can see why leaded fuel is on the way out. There are technical problems of reduced engine life when we remove all the lead. The auto makers of the world are working diligently to solve these problems.

Around 1966 there appeared a device known as the Air Injection Reactor (AIR) system, or air pump. This is an air pump powered by the engine via a belt drive. It takes power that costs gasoline. What it does is to inject air under slight pressure into the exhaust manifold. The engine exhaust includes unburned hydrocarbons and carbon monoxide, which would become pollutants or toxic agents if permitted out the tail pipe. As they leave the engine cylinders they are still very hot and would burn if mixed with oxygen. The air pump provides the oxygen required, and we rid the world of unburned hydrocarbons down to a level acceptable by prevailing standards.

Another ingenious device is termed the Exhaust Gas Recirculator (EGR). Take some of the exhaust gases and mix them with the air/fuel mixture before combustion. The result is a lowering of the peak combustion temperature, and this reduces the oxides of nitrogen (NOx) that come out the tail pipe. NOx are a prime ingredient of photochemical smog. The only problem from the driver's standpoint is that performance is noticeably reduced and one tends to use more pressure on the accelerator pedal to get moving, hence more fuel is consumed.

Woven into the emission equation is a spark curve that was re-

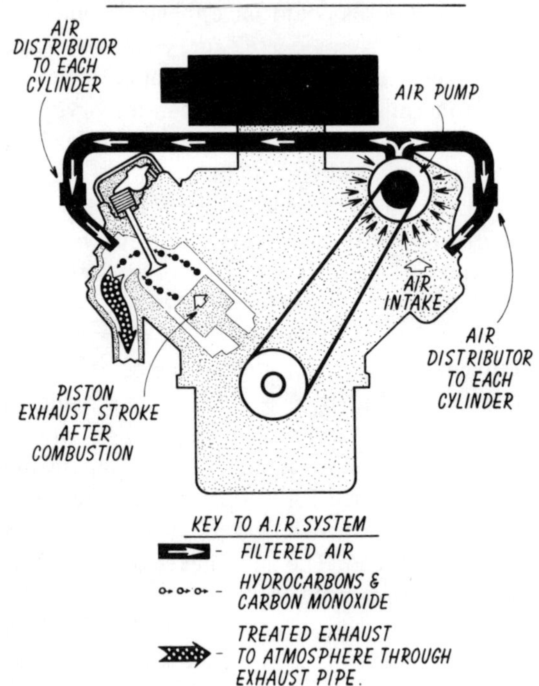
AIR INJECTION REACTOR (AIR)
AIR DISTRIBUTOR TO EACH CYLINDER
AIR PUMP
AIR INTAKE
AIR DISTRIBUTOR TO EACH CYLINDER
PISTON EXHAUST STROKE AFTER COMBUSTION
KEY TO A.I.R. SYSTEM
- FILTERED AIR
- HYDROCARBONS & CARBON MONOXIDE
- TREATED EXHAUST TO ATMOSPHERE THROUGH EXHAUST PIPE.

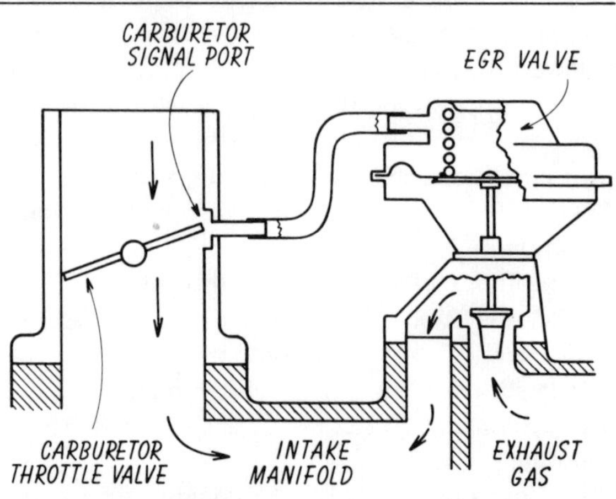
EXHAUST GAS RECIRCULATION (EGR) SYSTEM
CARBURETOR SIGNAL PORT
EGR VALVE
CARBURETOR THROTTLE VALVE
INTAKE MANIFOLD
EXHAUST GAS

tarded a few years ago. Emissions are cleaner but acceleration is not as good. As noted, there is more and longer pressure on the accelerator pedal to get the same level of performance with fewer miles per gallon.

Finally, automobile critics damn large engines as fuel wasters. In many cases they are, but most such engines were developed and installed to compensate for the lack of power that has come about because of emission controls. And we need the torque from these larger engines to operate the accessories many people seem to want.

It is a vicious circle that promotes inefficiency: Emission controls cause poor performance. To compensate, the driver presses a little harder and uses more fuel. Larger engines (as a factory compensation for the reduced performance) use more fuel. It is a difficult situation for the motorist who would like to get the most out of a gallon of gas.

Fortunately, in the interests of sanity, we have decided that we can live without quite as much power as we thought we needed. The thinking driver can get from A to B without setting speed records and, by following a few simple rules, come close to equaling the gas mileages common in the 1960s. (Of course, had he been interested in maximum economy during the 1960s, his gas mileage would have been phenomenal.)

The automobile industry is dynamic; no one knows when a new development may appear. It is clear that the industry's engineers are doing everything possible to develop efficient and economical vehicles. We know why mileage isn't so good, and until it is, it behooves every motorist to practice good driving habits and to do the best he can within the framework of what is possible.

WHICH GRADE OF FUEL?

You can save money and increase the fuel economy of your car by using the proper grade of gasoline. For years there was little choice; it was either regular or high test. The smaller workhorse-style engines all burned regular and the sporty high performance engines ran best on high test. The advent of low lead and unleaded gasolines in the interests of environmental protection has made a mystery for many out of which grade of gasoline to buy.

For many years lead additives have been used to increase the octane ratings in gasoline, and all engines were designed to run on a leaded compound, either regular or premium. If your older car uses a regular grade, be sure the gas has some lead content or you may do some damage to the valves. (Lead helps lubricate the valves.) Although it seems ridiculous to use top-grade fuel in a car that needs only an 80 octane gasoline, in some areas it is necessary because there is no leaded regular on sale.

If your older car has a high compression engine—9.5:1 or above—you just about have to use the highest and most expensive grade of gasoline. It is not economically feasible to detune a high compression engine so that it will operate well on unleaded gas. Beware of the absence of lead in the middle-range fuels.

All 1971 or later (domestic) automobiles are designed to run on low lead or unleaded regular gas. There is a slight catch here though. Each owner's manual today lists the type of fuel the car needs, and many manuals talk about a "91 Research Octane Number." By law the gas pump in your local station must have the research octane number posted on the pump. If your car needs the number 91, take care to check the low lead and unleaded octane numbers on the pump; many of them are posted at 89. It will take 91 or higher to satisfy the engine's octane requirements, so be sure the regular gas you are using meets the specifications. Otherwise you should go to the next higher grade of fuel or mix half and half between high test and regular in each tankful.

A good indicator of whether or not you are using the right grade of fuel is the car's performance. If you car develops a "running-on" habit, often called dieseling or post-ignition cough, you need to go up a grade in gas, assuming the engine is otherwise in good mechanical condition. The trait of running on after the ignition has been shut off is a symptom of poor combustion and should not be ignored or bigger and more expensive trouble may develop.

In many cases it is just as good for the budget to use a higher grade of gasoline. If your local regular is 89 octane, for example, and your car requires 91 or better, you will probably get more miles per gallon on high test than on regular. This seems to be especially true with four-cylinder subcompacts. Using the better grade of gas often will produce better fuel economy at the same cost per mile as you would have had with the cheaper gasoline.

If your owner's manual does not list the car's octane requirements, check with your dealer. The service department will have the octane information, and should also be able to relate it to the grades of fuel available in your area. The quality of gasoline varies with the seasons of the year and with the location. There can be no absolutely safe advice on the best overall fuel for any car.

HOW THE ELECTRICAL SYSTEM RELATES TO MILEAGE

The average driver may only become aware of his electrical system when it fails, as when the battery dies on a cold winter morning. At that point it is a bit late to prevent trouble, but a few facts may help in the future because the electrical system has a direct relationship to fuel economy. The driver should know that nearly all late model cars are equipped with alternators that produce the electrical energy to keep the battery charged. Hardly any manufacturer still uses the less efficient generator.

Electricity is energy. It is vital to the running of every automobile and there is only one source from which it can be ultimately produced—the fuel tank. Energy is required to turn the pulley on the alternator, and additional energy is required to operate every electrical device on an automobile from the turn signals to the electrically heated rear-window defroster.

Because we want to operate all the automobile's electrical systems at peak efficiency, there are some simple checks that should be performed on a regular basis. The battery electrolyte fluid should be kept at the recommended level. Most service stations check battery water regularly, but request an inspection about once a month if no one has bothered. It is also a good idea to have a mechanic check alternator belt tension every six months or so. Belts get old, worn, and stretched; a loose belt is going to slip, which is bad, and eventually it will break, which is very bad if the motorist is caught far from repair facilities.

The spark plugs, distributor, condenser, and ignition breaker points are part of the electrical system, and they have a direct effect on gasoline consumption. It is generally accepted that spark plugs should be cleaned and regapped or replaced with new ones when beyond service limits—about every 12,000 miles. Anytime a mechanic gets that far into the ignition system he might as well go all the way and reset the timing and check the points, replacing points as required.

Some new cars have switched to solid-state electronic ignitions, which do not have breaker points. This type will fire spark plugs in far worse condition than the old style ignitions, which means that less service is required. Now that the lead content has been drastically reduced in our fuel supply, spark plugs should be less susceptible to lead fouling. This could mean that the old 12,000 mile rule can be extended. Check your owner's manual or ask a good service facility for guidance.

VACUUM GAUGES—HOW AND WHY TO USE THEM

One of the best aids to teaching economy driving habits and maintaining them indefinitely is the vacuum gauge. It will not increase fuel economy in a direct sense, but if the driver will heed its simple indications it can have an excellent effect on gas mileage. A good one will cost about $25 installed.

What the vacuum gauge does is not difficult to understand. It measures air pressure in the intake manifold in inches of mercury and gives us some graphic clues about overall engine performance. (It is not necessary to understand about inches of mercury; just accept the numbers and what they mean from low to high.) The vacuum in the intake manifold decreases as the carburetor butterfly valve opens, which happens when the driver's foot presses the accelerator pedal. The vacuum also decreases with engine loads, as when an automobile climbs a hill, maintaining the same speed as it did on level ground.

In general, then, a high vacuum gauge reading indicates good economy, while a low reading means higher fuel consumption. It is not possible for a driver to try to maintain a set reading because automobiles vary. Most gauges are calibrated with a green arc in the midrange, and this means just what you would think; green is good. The red at the low end indicates that this is the area of the worst fuel economy. A white arc at the extreme high end is usually attainable only at idle.

Once a driver sees a vacuum gauge fluctuate in response to the pressure of his foot on the accelerator, a clear picture arises in his mind of where the fuel is going, and how driving habits might be modified for more miles per gallon. Just remember: high is good, low is bad.

A vacuum gauge also serves to provide a generalized picture of engine performance. One can obtain a clue about sticking valves from

VACUUM GAUGE CLUES TO ENGINE CONDITION

SOLID ARROWS REPRESENT STABLE NEEDLE CONDITION.
DOTTED ARROWS INDICATE NEEDLE MOVEMENT.

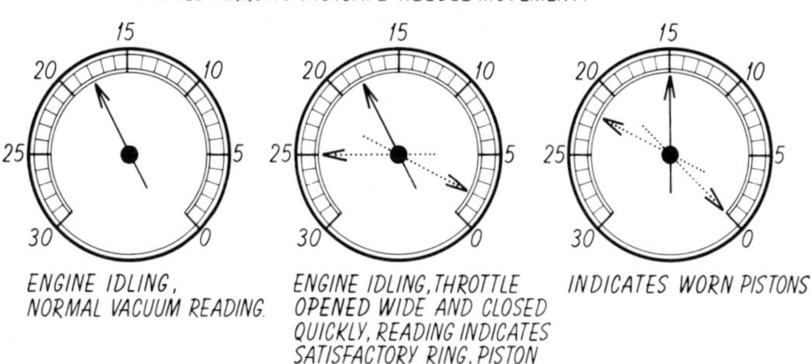

ENGINE IDLING, NORMAL VACUUM READING.

ENGINE IDLING, THROTTLE OPENED WIDE AND CLOSED QUICKLY, READING INDICATES SATISFACTORY RING, PISTON AND CYLINDER WALL CONDITION

INDICATES WORN PISTONS

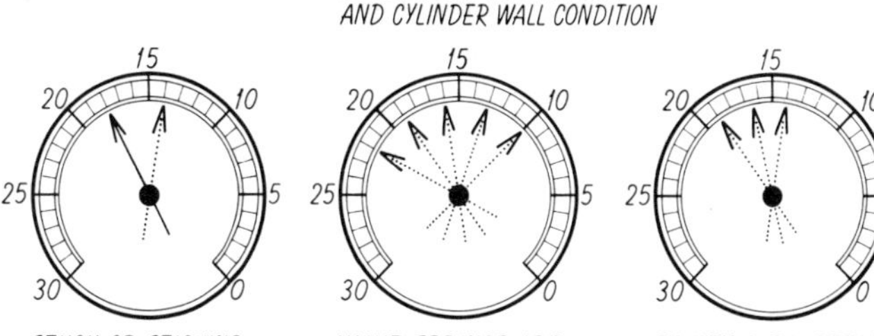

STUCK OR STICKING VALVES NEEDLE DROPS EACH TIME THE MAL-FUNCTIONING VALVE OPERATES

VALVE SPRINGS ARE WEAK OR BROKEN. BLOWN HEAD GASKET.

FAULTY CARBURETOR ADJUSTMENT. WORN VALVE GUIDES.

the needle movements, for example, and this will relate directly to fuel economy because an engine with a sticking valve is not delivering its best gas mileage.

THE FUEL FILTER

Where superclean fuel is critical, as in a fuel-injected aircraft engine, some means must be found to remove potentially harmful fuel impurities. And why not in an automobile? Fuel filters are readily available, and are just one more small step in the process of keeping the engine in first-class condition.

The reasons are very basic. First, there is always a certain amount of dirt in the bottom of a fuel tank, and it will tend to enter the engine if the fuel supply is run down near the bottom. Second, you can never be certain that gasoline as pumped is completely pure. Some oil companies have filters built into their pumps, and with this you should be safe. In most cases, however, there is no way to tell what is coming out of the gas hose.

The evidence appears when the filter is removed for its regular cleaning. Bits and pieces of dirt, metal, and corrosion are always in view, and the process is accelerated when operating in a dusty environment.

Some of the newer automobiles have a porous bronze filter built into the fuel line. It is necessary to clean this out on a regular basis, as outlined in the owner's manual. Accessory fuel filters, which cost under $5, are available, but remember that it is critical to install these devices so that the fuel flow is in the direction of the arrow on the filter. A canister filter is available, which is discarded when dirty and replaced in much the same way as an oil filter.

It is not possible to state flatly that the use of a fuel filter will improve gas mileage by a measurable amount, but preserving the engine's innards as a general policy will contribute to overall performance, and this will result in improved fuel economy.

16

Tuning and Maintenance

ECONOMY TUNE-UP TIPS

It wasn't very many years ago when there were all sorts of economy-tuning tricks that could be implemented. They worked, too, albeit sometimes with a slight sacrifice in performance. These tricks still exist but most of them are under a cloud of questionable legality. The problem is that they involve changing ignition distributor advance curves and making certain carburetor alterations to lean the fuel/air mixture—all changes in specifications that are liable to affect the exhaust emission standards as established for the car by the manufacturer.

The chances of an individual owner being caught making such modifications are not great. Actually not every state has the legal machinery to punish car owners who make these changes. However, in those states that do have laws against this, where there are regular or random inspections of emission control equipment, it will mean a citation for the owner who deviates from the manufacturer's settings.

What this means is that anyone who disconnects his emission control equipment, or does anything to make his engine operate more efficiently, is running afoul of the law. There are some exceptions; several states have lists of approved aftermarket engine accessories that have passed smog tests. The motorist who is inspected and found to have a set of approved exhaust headers, for example, will be okay, while the driver next in line with a nonapproved intake manifold will be in violation of the law.

Sometimes the system doesn't make much sense, and it frequently lags behind the realities of current technology, but it is the system and there is very little anyone can do about it.

We are not going to provide a do-it-yourself tune-up course here; there isn't enough space. For those drivers interested in home shop work—and this can save considerable money, along with providing the satisfaction of knowing that the work was done as it should be—it is suggested that the appropriate service manual be ordered from the car's manufacturer. Be prepared to buy a certain amount of test equipment; cars are significantly more complicated than they were ten years ago.

The best advice for the majority of car owners who neither desire nor have the ability to tinker under the hood is to read and understand the owner's manual, at least the part about recommended maintenance. Some of it may seem superfluous and a good way to enrich the local repair shop, but each operation is designed to lengthen the automobile's service life.

Pay particular attention to the regular servicing of emission control equipment because a malfunction of this equipment can have a bad effect on fuel economy.

The standard areas for regular engine maintenance still hold true: the electrical system including the spark plugs, the distributor points and timing, and the carburetion. Remember that such simple items as one misfiring spark plug can cut your fuel economy by 8 percent on a V-8 engine. Incidentally, on many of today's cars changing spark plugs can be a major task. There always seems to be a plug that is difficult to reach, and the inclination at some shops is to ignore it as if it would clean itself. It won't. The car owner who encounters such treatment, usually reported by another shop at a later date, would do well to forego giving further business to the offender. In fact, a service shop where the work is good and the prices fair is something to cherish. On the subject of prices, assuming that it is not necessary to replace the large parts of the carburetor, it should be possible to get an electrical and carburetor tune-up for about $50. Large shops with high overhead may find it necessary to charge more.

The fact is that if an owner will play the maintenance game as the factory intended, not cutting too many corners, he will find that his car will run as economically as it is supposed to. Higher gas mileage is, of course, up to the driver.

DON'T BE BILKED BY "FUEL-SAVING" GADGETS

You have seen the advertisements in magazines and perhaps even received mail-order solicitations. "Install our widget yourself in five minutes without tools, and instantly get blazing new power plus fantastic economy from your car—regardless of its age or condition." The price varies, but it's generally under $20. There is, of course, always a money-back guarantee. And the advertisement is filled with testimonials to the widget's efficacy.

Sounds great, doesn't it? Don't believe it. Someone once added up all the fuel-saving products listed in a mail-order catalog and decided that he should be able to get 200 miles per gallon if they all performed as advertised. They don't. That's not to say that somewhere there isn't an inventor with a $20 device that will double our gas mileage. But so far no one has found him. The man who develops such a product can write his own ticket at any of the large automobile companies. He won't have to bother with selling it by direct mail advertising.

Just so you can avoid being taken in (some of these sales pitches make the deal sound so good you cannot refuse), here are some typical examples that make the rounds. One gadget that does absolutely nothing is sometimes called a minisupercharger. It is certainly mini, but it is nothing more than a fan that spins as air is drawn in across it at the top of the carburetor. It doesn't supercharge anything, except the promoter's bankroll. Another device regulates the pressure of the fuel pump. It works, but if it is needed at all the pump is already in such poor mechanical condition that it should be repaired or replaced.

Higher performance spark plugs are frequently touted as more fully exploding the mixture in the cylinders. The fact is that nearly always these plugs are of poor quality and perform less efficiently than the nationally advertised conventional spark plugs.

One device that does help fuel economy is strictly a gimmick—it resists the driver's foot when he presses on the accelerator harder than he should. It is only a reminder. The prudent driver should himself be able to remember what to do, and save his money.

There are several attachments for injecting various fluids, including water, into the combustion mixture. The net effect here is to burn a leaner mixture, which will improve economy. On the other hand, there is a risk of damage to the valves.

Electronic accessories all sound so scientific that they must be good. In fact, most of them are useless. Capacitor discharge or transistorized ignition systems, which may cost upwards of $75, do work, however. They prolong the life of the electrical system, and may actually cause a fouled spark plug to fire where it would not with conventional ignition. Under these circumstances fuel economy will improve, but a motorist could achieve the same results with a normal electrical system tune-up—new plugs, distributor points, and the resetting of the timing.

The truth is that automobiles must obey well-known physical laws, and the most important law is that you do not get something for nothing. Remember the next time you are offered a device that looks as though it should double your gas mileage, save your money for legitimate work from recognized shops.

THE BEST TIME OF DAY TO FILL UP

It is worth considering when is the best time of day to fill the gas tank. Fill up in the morning or the evening, or whatever is the coolest part of the day; it can make a small difference in the amount of usable gasoline that goes into the tank. The colder the gas flowing into the tank, the more gas will fit into the tank. As the fuel temperature goes up the gas expands, requiring more room, so the colder the gas the more combustion per gallon. Fuel temperature makes a difference in accurately measuring miles per gallon. In economy-run competitions the gas hose and tank temperatures are measured and corrections are made to compensate for temperature differences between various cars. For example, tests on a special gas tank in the trunk of a dark-colored car have shown that it will have a slightly higher fuel temperature on a sunny day than the gas contained in a similar tank in a white car.

The best savings in fuel can be obtained by stopping for gas when the station is uncrowded. Idling along in a line approaching the pump is wasteful, and the best plan is to avoid rush hours.

On a long trip plan on filling the tank each evening. Next morning you can motor past the crowded urban stations and head for the open road.

Of course with current gasoline shortages, mandatory rationing plans, and reduced hours of service station operation, it may not be possible for some time to come to stop for fuel just when you want to, or even to obtain a full tank of gas when you do get to the pumps. But whatever standards and regulations are established in the future, the above advice is sound and should be heeded if it is possible to do so.

WHY KEEP THE TANK FULL?

The fuel crunch has encouraged motorists to keep their tanks full at all times. It is downright crazy to drive from station to station, getting a few gallons each time, wasting time and fuel in the search, but in areas where fuel is in short supply that is the way it is. In addition to being sure that the tank has fuel, there are other good reasons for keeping half a tank or more on board all the time.

The first reason is the importance of keeping at least a day's supply of fuel for emergencies. Although evaporation has been virtually eliminated with the 1974 cars, older cars are subject to it, and there will be less evaporation from a full tank.

Be careful filling the tank in warm weather. It is wise to plan to drive a bit after topping off, at least enough to burn the fuel down from the neck, because expansion from the sun's heat will cause a completely full tank to overflow.

The exception to the half tank rule at all times is during highway travel. It makes little sense to make frequent stops to top off the tank. Plan the trip, estimating fuel needs, and stop for gasoline at about the one-quarter mark on the gauge. A quarter tank provides enough reserve to hit a gas stop except in the most remote regions.

INSPECT FOR FUEL LEAKS

It would seem that no one interested in saving fuel would let even small amounts of gas leak away. Yet it happens, and if the leak is sufficiently large it can pose a serious safety hazard along with the obvious detriment to fuel economy.

Fortunately the best way to check for external leaks is simply by looking. Although a thorough examination is a job for a qualified mechanic, the owner can make a reasonable inspection without any special knowledge.

The first step is to raise the hood and look for fuel stains, starting at the carburetor and working back through the fuel lines. It may be necessary to remove the air cleaner to get a good look at the carburetor. If fuel stains are present, determine if they are fresh or due to old seepage by wiping with a rag. Should there be an obvious seepage of fuel, and there is the presence of fumes, there is a leak. Unless the owner is pretty good with a set of tools, it will be best to let a professional mechanic fix it.

The rest of the inspection is best made with the car on a hoist, where the tank and its various fittings can be examined. A leaky tank seam is serious; it may be necessary to remove the tank from the car and have it welded, once again a job for a man who knows how to handle welding equipment safely around a fuel tank.

It is not possible to say how much fuel is likely to be lost from a fuel system because of minor or major leaks. On the assumption that we are trying to consume all the fuel we buy in the engine, as well as the fact that leaks pose a fire hazard, a leak should be corrected.

FUEL AND OIL ADDITIVES: YES OR NO

In this age of miracle chemicals there is a great temptation to believe that the addition of a can of some mysterious but widely advertised substance to the fuel will result in miraculous performance, longevity, and economy.

We are not prepared to argue the overall good and bad of oil and fuel additives. Many motorists swear by their merits. However, the fact that none of the major auto makers recommend their use is a powerful argument against them. What we can say is that there is no evidence that they contribute to increased fuel economy. Obviously, fuel additives burn but they offer a costly method of increasing the supply in the tank.

There is no such thing as a tune-up in a can, in spite of advertising claims, and we have never been able to measure any improvement in fuel economy from the use of any additive.

MAINTAINING YOUR TIRES FOR ECONOMY

There is no secret to keeping tires properly maintained, and it isn't even very difficult. Inflation, alignment, balance, rotation, and inspection are the things to be concerned about. Should any of the first three be abnormal, there will be two unwanted results: poor tire wear and reduced fuel economy, because the tire will have more rolling resistance than it should. (There is one exception that will be discussed concerning overinflation and economy.)

Don't mix tires of different types. In other words, don't use both bias-ply, bias-belt, and radials. All four tires should be of the same type or the car may not handle as well as it should.

Inflation is easy. Use the recommended pressures, check the tires when cold about once a month, and invest in a good tire gauge. Service station gauges are not always accurate. Most new cars have two pressure recommendations, the higher one being for maximum load. It won't hurt to run at the higher pressure, and it will lessen rolling resistance. Incidentally, tires checked when cold will build 3 or 4 pounds of pressure after a very modest heat buildup. Don't remove this air; it is not excess.

It is not a good idea to permit a tire to run underinflated; it will wear excessively near the edges of the tread, and rolling resistance will be increased, hence fuel economy suffers.

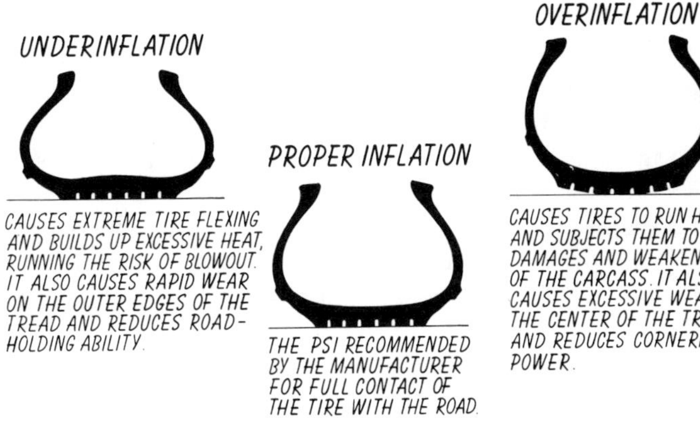

On the other hand, overinflation will reduce rolling resistance and improve gas mileage. In gas mileage competition drivers have been known to pump tires to rocklike hardness in order to minimize rolling resistance. Pumping the tires beyond the manufacturer's recommendations, however, is impractical for several reasons. The first is excessive wear on the center of the tread. We have observed overinflated tires worn out in the center and rendered unserviceable, while still having a perfectly good tread along the edges. It is a waste of good rubber. Second, the quality of the ride will suffer when tires have too much air, and handling may be adversely affected. The gain in economy is more than offset by these factors.

Alignment and balance are not items that a driver can correct. Have a good station check both items once a year. Symptoms of poor alignment are odd wear patterns on the front tires. A good mechanic

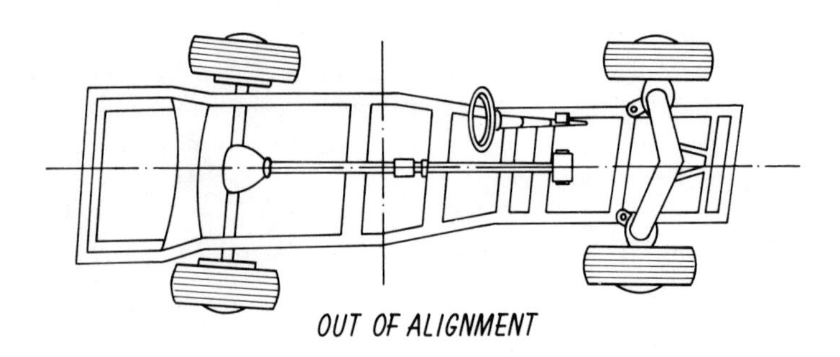

OUT OF ALIGNMENT

can verify bad wear in a moment. Tires that are out of balance often can be felt by the driver as a thumping noise is heard at certain speeds. In really bad cases the tire will show wear at the point of imbalance.

Most tire manufacturers recommend the regular rotation of tires to equalize wear. Switching at 8,000 mile intervals seems to be about right. There is one school of thought that says that it is not necessary to rotate radial tires. But even here, the advantage of rotation when it includes the spare is that five tires will provide 20 percent more tire life than will four.

A visual inspection should be performed on a regular basis, either by the owner or a tire shop. You are looking for cuts, breaks, or signs of unusual wear.

No one can say absolutely that proper tire maintenance will boost fuel economy by a certain amount. It is really a matter of improper maintenance reducing gas mileage. We do not want that to happen, so the best preventive is to observe these few simple rules for tires.

CHECKING EMISSION CONTROLS FOR BEST OPERATION

Emission control systems are going to be around for a long time. In

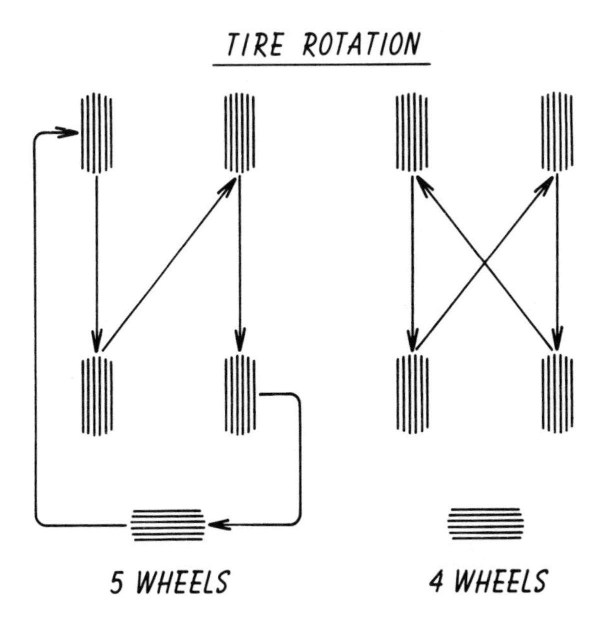

terms of fuel economy we would rather not have most of them, but in most states it is against the law for an owner to disable them, and doing so would be contrary to a reasonable approach to the air pollution problem. The fact that low emission engines take more fuel than the high emission engines of a few years ago is something we must live with. And living with it means that the various emission control systems must be as efficient as possible. Most of them are not prone to failure but a malfunction at any level opens the door to reduced fuel economy at a time when it is already lower than we would like.

These systems may have slightly different names within the automobile industry but they all do the same thing, which is to reduce unburned hydrocarbons, carbon monoxide, and oxides of nitrogen. Not all cars will have each system. It depends on the engine and year of manufacture.

Positive Crankcase Ventilation (PCV) is the oldest system. In most situations it helps improve gas mileage. During the combustion process a highly corrosive gas is produced that, along with water, leaks into the crankcase. Called blow-by, it must be removed from the crankcase before it reacts with the oil to form harmful sludge. Also present is unburned fuel, which will dilute the oil if not removed.

The PCV system uses engine vacuum to draw fresh air through the crankcase to remove these harmful gases. The PCV valve varies the amount of airflow through the system, depending on the speed of the engine. The vapor is returned to the intake manifold and is combustible, and this has been shown to provide better fuel economy over the old method of venting engine blow-by out the exhaust pipe. At idle, for example, fuel economy with the PCV is 15.4 percent better, while at 50 mph the gain is about 2.3 percent.

All one need do is follow the manufacturer's recommendations for servicing the PCV system—generally cleaning it every 12,000 miles and replacing it at 24,000 miles.

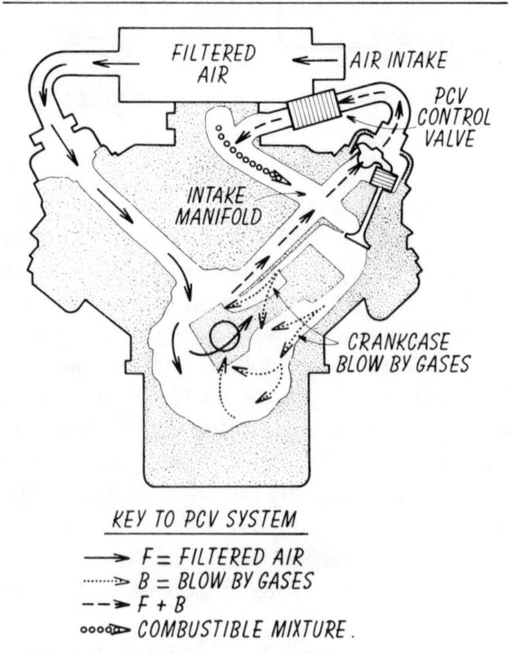

The Evaporation Control System (ECS) has one function: to reduce fuel vapor emissions that normally vent to the atmosphere from the gas tank and carburetor fuel bowl. It is part of what has made the modern car a plumber's nightmare. There are changes in

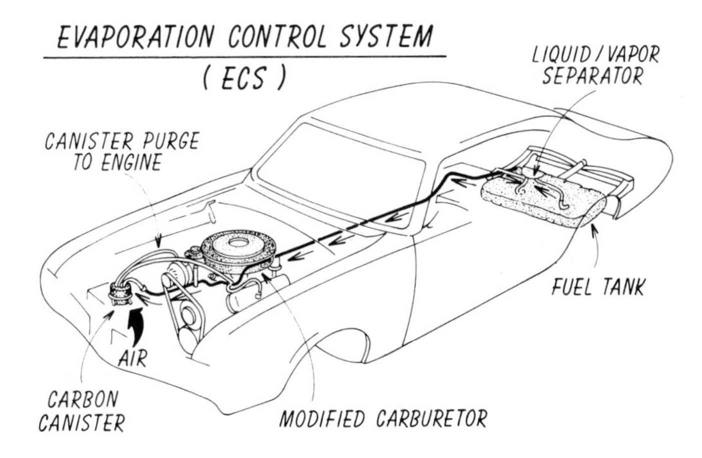

the carburetor, the fuel tank, and the tank venting system so that vaporized gasoline is collected and directed through a liquid vapor separator to a canister of activated charcoal. Fuel vapors attach themselves to the charcoal when the vehicle is at rest. When it is driven the canister purge system is activated; the engine vacuum draws air through the canister and picks up the fuel vapor, which is taken into the induction system and burned. When the canister reaches a saturation point, the excess vapor is allowed to escape.

A system malfunction may cause poor idling or loss of fuel. The owner can check to see if the hoses to the canister are attached correctly and not pinched. However, it takes a shop examination to check the canister itself, to change its filter, and in some cases inspect around the canister with an infrared tester to detect the presence of hydrocarbons.

The biggest enemy of fuel economy is the Exhaust Gas Recirculation (EGR) system. Its purpose is to reduce oxides of nitrogen (NOx) from the exhaust fumes. NOx formation takes place at very high temperatures, that is, during the peak temperature period of the combustion process. To control NOx formation, engine temperatures are lowered by introducing small amounts of an inert gas into the combustion process; the end products of combustion provide a continuous supply of relatively inert gases and it is just a matter of using those gases in the correct proportion.

In order to dilute the intake mixture with up to 15 percent of the exhaust, additional exhaust passages are cast into the intake manifold. The key to the system is a vacuum operated shut-off and metering valve, the EGR valve. The system is not required at idle, and maximum flow occurs at 30 to 70 mph cruise speeds.

The EGR valve can fail, as may the thermal vacuum switch, also part of the EGR system. Unfortunately, complete servicing is not for the amateur, but the owner may inspect to be certain that the vacuum line is properly attached to the EGR valve.

A great many cars with manual transmissions use a device called the Transmission Controlled Spark (TCS) system. It controls the exhaust emissions by preventing ignition vacuum advance when the vehicle is operated in reverse, neutral, or low forward gears. In this system there is a vacuum advance solenoid, two switches, and a time relay. Because the spark is retarded during several driving conditions, performance suffers and the driver tends to use more fuel to compensate for this reduced performance. Vacuum advance is provided in top gear, at low temperatures, and the time relay energizes the solenoid for advance 20 seconds after the ignition key is turned on. This makes it easier to keep the engine running after a cold start.

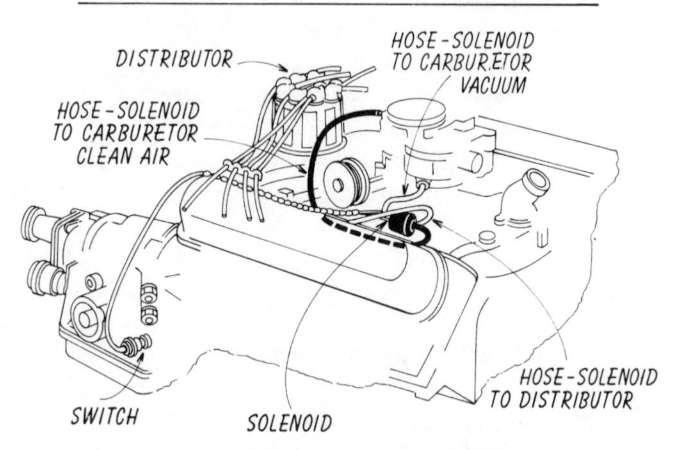

Although the system is not overly complex, any of the two solenoids, the two switches, or the time relay could malfunction, causing stalling, running on, high idle, poor high gear performance, deceleration exhaust "pop," and—worst of all—excessive fuel consumption. Troubleshooting and repair requires a skilled mechanic with test equipment and extensive knowledge of the system.

The Air Injector Reactor (AIR), already discussed briefly, is one of those devices that reduces the emissions of hydrocarbons and carbon monoxide when it is properly installed and maintained. However, if any AIR component or any engine component that operates in conjunction with the AIR system should malfunction, the exhaust emissions may increase. Fuel economy certainly won't be helped by any failures, and economy is not aided by the pump itself, which requires energy from the engine to turn.

All the motorist can do is to make a visual inspection of the air pump for obviously broken or missing parts and a broken, loose, or frayed belt. A service station will be the best place to obtain proper belt tension (or a new belt), replacement of the pump filter, find hose leaks, and examine the diverter valve. The worst symptom from the pump is going to be excessive noise—chirping, rumbling, or knocking. A defective AIR pump cannot cause poor idling or performance.

From these capsule descriptions and operating tips it should be clear that modern automobiles are literally jungles of emission control equipment, most of it too complex for an owner's easy diagnosis. Not every car has every system, and there are some new systems on the horizon. Hopefully, the coming years will see the development of engines with low enough emissions so that many of today's devices can be forgotten. Perhaps then we can get back to fuel economy as it existed in the pre-emission control days.

CHASSIS MAINTENANCE FOR ECONOMY

If we examine the chassis as it relates to economy, the name of the game is to reduce rolling resistance. We do not want any obstacle in the path of getting the maximum horsepower from the engine to the wheels. Approached in this fashion, the areas under consideration become readily apparent: wheel alignment, brakes, and lubrication. All are items that should be in order for normal driving, but sometimes they are overlooked. The economy-conscious driver will have his vehicle examined during a regular maintenance check with special emphasis on these areas.

Wheel alignment is not a home shop job. It requires special equipment under the control of a technician. The toe-in or toe-out, and caster and camber should be set to manufacturer's specifications, and automobiles with independent rear suspension (such as rear-engined cars) will need the rear wheel alignment adjusted too.

Clearly it is a safety consideration to have brakes in good working order. But how often does a motorist have his car put on a hoist and then spin his wheels to test for brake drag, Brake drag, even a slight amount, wastes fuel and wears the brakes.

It should go without saying that areas designated for chassis lubrication should receive lubricants at factory specified intervals. Improper or neglected lubrication won't have a great effect on fuel economy, at least not for very long, because there will be other more serious consequences. For example, a wheel bearing without grease very quickly makes the vehicle unusable.

Take care of the mechanical equipment in your car and it will take care of you. It is almost a cliché, but it is true. An automobile with its chassis properly aligned and lubricated, with everything turning freely, will deliver long life and the best fuel economy.

Index